Urban Subversion
and the Creative City

This book provides a comprehensive critique of the current Creative City paradigm, with a capital 'C', and argues for a creative city with a small 'c' via a theoretical exploration of urban subversion.

The book argues that the Creative City (with a capital 'C') is a systemic requirement of neoliberal capitalist urban development and part of the wider policy framework of 'creativity' that includes the creative industries and the creative class, and also has inequalities and injustices in-built. The book argues that the Creative City does stimulate creativity, but through a reaction to it, not as part of it. Creative City policies speak of having mechanisms to stimulate individual, collective or civic creativity, yet through a theoretical exploration of urban subversion, the book argues that to be 'truly' creative is to be radically different from those creative practices that the Creative City caters for. Moreover, the book analyses the role that urban subversion and subcultures have in the contemporary city in challenging the dominant political economic hegemony of urban creativity. Creative activities of people from cities all over the world are discussed and critically analysed to highlight how urban creativity has become co-opted for political and economic goals, but through a radical reconceptualisation of what creativity is that includes urban subversion, we can begin to realise a creative city (with a small 'c').

Oli Mould is Lecturer in Human Geography at Royal Holloway, University of London.

'*Urban Subversion and the Creative City* is Oli Mould's bold plea for truly creative urban thought and action … Creative, that is, in a wide range of subversive but always social ways, and not only outside but against the softly suffocating hegemony of authorised versions of the Creative-Cities script, in all its banal ubiquity. Not before time, this is the creative city turned upside down.'

Jamie Peck, *Canada Research Chair in Urban & Regional Political Economy and Professor of Geography, University of British Columbia, Canada.*

'The Creative City of neoliberal urbanism has met its match. In this fascinating, meticulously researched, truly global book, Oli Mould introduces us to the creative city of subversion and desire. A much needed act of liberation from the official terrain occupied by the creative class.'

Professor Roger Keil, *Faculty of Environmental Studies, York University, Canada.*

'Oli Mould is a legendary scrapbooker of all things urban, and the city itself – its benches, billboards, pavements, parks – is his raw material. This book is the product of many years of quiet but critically discerning rambling through diverse urban places and through eclectic academic ideas. Mould has painstakingly traversed both the world's iconic and ignored cities, collecting and photographing, asking questions and talking with those who make cities genuinely creative places – not the cashed-up hipsters, Creative City policy wonks or think-tank 'expert' authors of glossy strategic plans, but the activists, artists and ordinary people on the street whose desire to use the city to other ends make meaningful life out of concrete, brick and bitumen. This book is a rare insight into how to think the creative city differently, a book that crackles with conflicting voices, but always sees the city as a space of possibility.'

Chris Gibson, *Director, Global Challenges Program and Professor in Human Geography, University of Wollongong, Australia*

Routledge critical studies in urbanism and the city

This series offers a forum for cutting-edge and original research that explores different aspects of the city. Titles within this series critically engage with, question and challenge contemporary theory and concepts to extend current debates and pave the way for new critical perspectives on the city. This series explores a range of social, political, economic, cultural and spatial concepts, offering innovative and vibrant contributions, international perspectives and interdisciplinary engagements with the city from across the social sciences and humanities.

Published

Urban Subversion and the Creative City
By Oli Mould

Urban Subversion
and the Creative City

Oli Mould

Routledge
Taylor & Francis Group

LONDON AND NEW YORK

First published in paperback 2017

First published 2015
by Routledge
2 Park Square, Milton Park, Abingdon, Oxon OX14 4RN

and by Routledge
711 Third Avenue, New York, NY 10017

Routledge is an imprint of the Taylor & Francis Group, an informa business

British Library Cataloguing in Publication Data
A catalogue record for this book is available from the British Library

Library of Congress Cataloging in Publication Data
Mould, Oliver.
    Urban subversion and the creative city / by Oliver Mould.
        pages cm. – (Routledge studies in urbanism and the city)
    Includes bibliographical references.
    1. Urban policy. 2. Cultural policy. 3. City planning. 4. Arts–Economic
aspects. 5. Cultural industries. 6. Urban economics. I. Title.
    HT153.M68 2015
    307.76-dc23
                        2014040875

ISBN: 978-1-138-79704-8 (hbk)
ISBN: 978-1-138-69328-9 (pbk)
ISBN: 978-1-315-75746-9 (ebk)

Typeset in Times New Roman
by Cenveo Publisher Services

Contents

Figures

Acknowledgements

As the cliché goes, writing a book is never a singular exercise, it involves the work and effort of a whole collective of people. This book is no different, and so there are a few people to whom I need to give my heart-felt thanks. First, there is the myriad of people who let me into their subversive and creative worlds. But specifically, Charlie, the young traceur from Guildford who opened the doors to the London scene for me in 2007, Moses Gates who explored with me first and asked questions later, and the members of the Long Live Southbank campaign group (Henry Edwards-Wood and Paul Richards in particular). You can't move history after all.

Second, my colleagues at Royal Holloway who provided encouragement, space and intellectual fodder for the book. In particular Katie Willis, Innes Keighren and Veronica Della Dora for their advice and guidance. Special mention though has to go to Harriet Hawkins, who not only put up with my verbal ramblings, but also gave constructive advice on the written output as well.

Finally, my wife Sarah was a continual source of support throughout the writing process. Listening to me whinge, moan and laugh, she provided all the emotional support that I needed and then some more on top of that. She gave me the space and time to work and never complained or deflected my attention in any way. This book would not be complete without her, nor would I.

Prologue

The man upon a wire

On 7 of August 1974, at around 7.15 a.m. Eastern Standard Time in New York City, Phillipe Petit stepped off the roof of the South Tower of the World Trade Center and onto a wire no more than three quarters of an inch in width. The wire was suspended between the two towers, a distance of 140 feet, and 1,368 feet above street level (Lichtenstein 1974). For approximately 45 minutes, Petit, a French circus performer specialising in tightrope walking, made his way between the two towers, performing his circus tricks along the way. He lay down on the wire, waved to the gathering crowds below, stood one legged, and even jumped from one foot to the other. He went back and forth eight times in total and even had the temerity to taunt the awaiting police officers on the roof. It was only the threat of an approaching helicopter that caused him to eventually give himself up to the police back on the roof of the South Tower.

The 'Man on Wire', as he became known, spent years planning this act. Along with his lifelong friend Jean-Louis Blondeau, they schemed, planned and meticulously plotted their 'attack' on the towers. Petit had already tightrope walked between the twin towers of Notre Dame Cathedral in Paris and the pylons of the Sydney Harbour Bridge, but the World Trade Center in New York City presented for Petit his ultimate challenge. In the months leading up to the event, he assembled a group of like-minded individuals from within New York and as a group they infiltrated the towers in inventive, audacious and subversive ways. They pretended to be French journalists in order to walk around the towers beforehand, photographing the intricate details of the building. They produced fabricated identification cards for a fake utilities company in order to gain access to the building with all their necessary equipment. They even secured the help of an insider. On the night before the stunt, Petit and Jean-Louis along with two other accomplices spend hours hiding from security guards in some of the most painful and uncomfortable positions imaginable. Atop the building, without any sleep, they somehow manage to achieve their objective of suspending the wire between the two buildings using nothing but a bow and arrow and their own naked bodies to feel for the wire in the darkness. In managing to pull off this incredible urban stunt completely covertly, Petit secured worldwide infamy (Petit 2008).

The enormity of the preparation and the spectacle of the act itself are documented in the film *Man on Wire* (2008) and accompanying book. In the film, the still images of Petit performing his high-wire act cannot fail to induce awe at how courageous and brazenly death-defying his actions were. However, what is also apparent from the film and the monumental efforts that went into, not only the act itself, but also the preparation, is the functional malleability of the buildings themselves. By performing a spectacular stunt atop two skyscrapers designed for nothing of the sort, Petit reappropriated them into a space of performance, of play, of circus-like dramaturgy. From that point on, even after their tragic demolition in 2001, New York City's Twin Towers will forever have an association with his daring act. What is more, Petit claims that the pursuit of his goal with such alacrity and veracity (to the point of accusations by his friends of insanity) was not to achieve the aforementioned reappropriation of the heart of one of the world's prominent financial centres into a space of subversive performance. Instead, he claims that he simply had a desire to do it because of his very nature as a high-wire performer – he had a desire to produce this performance. Moreover, he instantly saw those two towers as designed for him:

> Then I lean over the edge, ready to climb down the inclined columns to the six-inch ledge 11 feet below that connects the 110th floor with the 1,350 feet of verticality, so I can look straight down. I do not. Because that's when it strikes me: teeth clenched, eyes half closed, in horror, in delight, I manage to whisper my first thought (whisper, so the demons won't hear): 'I know it's impossible. But I know I'll do it!' At that instant, *the towers became 'my towers'*. (Petit 2008: 17, my emphasis)

This anecdote illustrates how creative engagement with the city can transform the built environment. There are countless events, perhaps not as spectacular as Petit's but no less transformative, happening day by day in cities all over the world. People are using their city and the built environment within them innovatively, playfully and creatively. Petit's act is the ultimate urban subversion of modern times (although the pair of urban explorers who scaled and 'dangled off' the unfinished Shanghai Tower in February 2014 come close for sheer visual spectacle (Metcalfe 2014)). Yet his purpose was not to subvert the urban environment. It was not a political act of defiance against financial greed. It was not an infiltration or tactical manoeuvre against the prevailing official urban agenda. It was an act of pure *desire*. His act may have been politicised by the myriad of subsequent deconstructionism texts (see Mackay 2011). However, for approximately 45 minutes on 7 August 1974, the World Trade Center became more than an office building in Manhattan, it became a high-wire performance, not through the sanctioned legitimisation of a corporate entity but by the pure will and desire of one man and his group of friends to perform this act.

1 Introduction

Sitting in a cafe on King Street, Newton in Sydney, I spy a postcard holder affixed to the opposite wall (see Figure 1.1). The holder's header reads, 'What creative life do you want for Sydney?' Within the holder sit a number of postcards and flyers advertising art galleries, operas and theatre productions throughout Sydney, all of which were either sponsored, publicised or hosted by the City of Sydney, the city's governing body. As can be seen in Figure 1.1, next to the esoteric question 'What creative life do you want for Sydney?' is the orange circular logo for the council and the 'Creative City Sydney' brand.

The brand is an initiative set up by the City of Sydney to use the 'ongoing commitment to culture and creativity' in order to, among other things, 'affirm the centrality of the arts and creativity' to the lives and economies of the city and 'revitalize the Oxford Street Prescient [one of the main commercial streets in inner-city Sydney]' (City of Sydney 2013a: n.p.). By deliberately enacting a number of policies aimed at encouraging artistic practice and creative industry activity and beautifying public spaces, the City Council has redeveloped a number of derelict buildings, fostered creative enterprise and staged citywide events celebrating the vibrant arts and cultural activity of the city. All these have been done as targeted activities that fall under the rubric of 'Creative Sydney', a suite of governmental actions specifically designed to aid urban development through the notion of 'creativity'. Sydney then, like a multitude of metropolitan areas all over the world, is very much a Creative City, with a capital 'C'.

Do you live in one of these Creative Cities? You may not be aware of it, but given the popularity of the 'Creative City' idea, there is a high probability that your current home city or town has enacted or has been developing policies designed to encourage primarily economic growth, but also social cohesion and cultural participation through the premise of 'creativity'. Since the turn of the new millennium, many urban and regional governments from all over the world, from Sydney to Sheffield, from Manila to Madison, have been enabling strategies of development purporting to stimulate creativity among its inhabitants and, perhaps more predominantly, to bring in talented, educated and creative people from elsewhere in the hope of benefiting from their economic growth potential. Moreover, gargantuan financial sums have been spent on these endeavours. Large-scale creative economy infrastructural developments have

Figure 1.1 Postcard holder in Newtown, Sydney.
Source: Author's photo, April 2013.

been commissioned; many, many cultural institutions have been built; artistic events have been staged and sponsored; urban infrastructure has been upgraded and beautified; financial incentives have been offered for the relocation of creative individuals, projects and firms; marketing and branding offensives have been launched; research and development exercises have been given the green light; promotional campaigns have been initiated – all in the name making a city more 'creative'. Indeed, the term 'Creative City' is now firmly entrenched in the parlance of urban politics. This is because, in the era where cities are run, managed and operated like corporations, they are engaging in competition with each other for the global flows of capital and, increasingly, for footloose talented and creative people. By effectuating a creative persona through a suite of policies, cities are looking to gain that vital competitive edge that will capture the people, creativity and, importantly, the economic rewards that they promise. Moreover, cities encourage these policies in the belief that the economic stimulus that comes from the presence of creativity diffuses throughout the city. By being 'creative', the city is more productive and innovative, thereby stimulating further economic rewards and creating a more attractive place for businesses and tourists to come to. In addition, being a Creative City (with a capital 'C') is purported to encourage social inclusion, cultural participation, poverty alleviation and housing stock renewal and generally creates a better place to live. The prosperous utopian idyll of a Creative City is indeed an intoxicating vision for many urban governments.

However, what does it mean to be 'Creative'? How do we define creativity in this context? Moreover, how do you relate such an innate human affective quality to a city? How can an entire urban environment actually be creative? These questions are often at the forefront of academic (and to a lesser extent political) debates about the 'Creative City' as a paradigm, yet cities continue to adopt such policies and agendas without any definitive and rigorously obtained answers. In the race to become a Creative City and benefit from all that which 'creativity' can bring, cities are actually far from stimulating the productivity and innovative capacity of their inhabitants (as the lexicon would suggest), they are distancing themselves from realising such a vision. Rather than fermenting a ubiquitous upscaling of the quality of urban life, the Creative City in all its various guises has engendered and crystallised long-standing and existing urban conditions of economic inequality, social exclusion, cultural desertification and disengagement, and in some case, social and political unrest. As such, there are many dissenting voices regarding the Creative City paradigm, and how it is being enacted in contemporary cities. Accused of fuelling the gentrification process, many of the schemes that are formulated under the rubric of 'making a city more creative' are criticised for being just a policy 'fad' – simply the latest justification for an already ongoing process of economic urbanisation characterised by a privately led investment ethos that prioritises financial reward over any other social or cultural mandate. The associated creativity paradigms of the creative class, the creative economy and the creative industries, which contribute to the economically deterministic qualities of the Creative City paradigm, have also received similar critical rebuttal, given their political, economic and therefore systemic inextricability from the Creative City ideology. The political mobility and inherent 'ease' of applying this ideology means that it spreads from city to city, copying the perceived successful political and economic 'model' rather than engaging with the different set of localities, histories, cultures and social issues.

As such, the major *problem* of many of the policies embroiled within the Creative City rhetoric is that they have a deleterious effect on the social fabric and cultural diversity of cities, and have very little to do with affecting creativity in individuals at all. Moreover, such policies are designed to be replicable, to be globally mobile and reproduce homogenous urban spaces internationally. Put bluntly, they are actually the *antithesis* of creativity. The Creative City evangelists have hijacked the *idea* of creativity and turned it into an economically deterministic pastiche of what it means to be creative, packaged it and sold it to the world. Furthermore, embroiled within this process is the total *reconceptualisation* of 'true' urban creative practices – those of reappropriation, transgression, interventionism, subversion, resistance, activism and experimentation – into oppositional activities to be resisted or capitalised upon. The Creative City paradigm therefore, while being the latest iteration of an continual, politically systemic, gentrifying, capitalistic and urbanising policy discourse, is also (re)producing alternative creative practices (with a small 'c') as 'the other'. These practices of urban subversion (as I have come to call them) have an iterative history, but are proliferating and becoming more 'creative' in reaction to (and between the cracks of), rather than in conjunction with, the contemporary Creative City ideology. These modern-day creative practices (that have a long, political and social urban history) can be identified and perhaps most accessibly articulated as creative urban subcultures, the more recognisable including parkour, street art, graffiti, skateboarding, urban exploration, yarn-bombing, buildering and flashmobbing, each with their own internal and interdisciplinary tensions. Yet, beyond such subcultures (or identifiable grouping of 'subversive' cultural and creative practices), there is a less teleological and more fluid, rhizomatic and experimental universe of everyday urban practices, where people are simply reconfiguring the city around them to express their cultures, beliefs, anxieties, frustrations, happiness and a whole range of other affective and emotional occurrences. The homeless ex-serviceman who chalked anti-war and biblical scenes on the Camden Town pavement on which he slept; the man who bought a tank and used it to voice his discontent at the local council; the little girl who decides to ignore the 'do not climb' sign to play on public art; the disabled man who built his own ramp to the Municipal Health Building in Juína as he was dismayed at a lack of 'official' action; and as the prologue recounted, the Frenchman who tightrope-walked across the Twin Towers in New York – all these instances are of people changing the city around them for their own desires (albeit with some more spectacular results than others). They are being *creative*. They are not led by economic and/or urban development requirements or by state-level political motivations (although, as will become apparent in this book, these may be subsequently inferred in the contextualisation and representation of their practices). They are inherently part of an affective, non-representational urban fabric of creative sociality. They are fundamentally being active urban citizens rather than passive consumers of a spoon-fed 'Creative' urbanism. However, within this broad idea of urban subversion, their resistance to, or indeed their inclusion in, the Creative City paradigm is a highly contested and fluid dialectic. This is because Creative City policies themselves are looking to such activities in order to maintain the uniqueness and innovative image that propels economic competitiveness, in other words to maintain their brand. Therefore we see commercialised forms of parkour and skateboarding, urban exploration tours of peri-legal sites, community-led planning interventions being sponsored by city governments and so on. But in so doing, the Creative City, despite rhetoric and 'spin' to the contrary, is ultimately reducing such activities to economically determined instruments of urban development and politically, conceptually and linguistically whitewashing any transgressional, subversive or resistive characteristics in favour of more putative urban and economic development aims that can be homogenised and replicated. And thus, returning to the original *problem* – the Creative City as the antithesis of urban creativity.

This book outlines how we can begin to disrupt this process. Because *being* creative is a critical attribute to *being* a city, so the notion is worth rescuing from the problems that plague it. Hence, through a vehement excavation of the Creative

City paradigm and its various political vernaculars (namely the creative industries, the creative economy, the creative class and associated and incumbent lexicon), this book will analyse how the idea relates to the growth of urban entrepreneurialism more broadly and its associated deleterious impact upon the social and cultural fabric of cities and desire for homogenous urbanity. More than this though (as such critiques have been well-rehearsed in the last decade or so of critical urban studies literature), the book will theoretically explore how urban subversion has proliferated contemporaneously with(in) the Creative City, and detail the complex interaction between these idioms which are at the same time contested and related, present and absent, near and far, dichotomous and intermeshed, democratic and hegemonic, major and minor. By highlighting the economic instrumentalism inherent in the Creative City and positing it against an urban creative social fabric that is simply the *desire* to reconfigure the urban environment by active citizens, this book exposes the current Creative City idea to be too instrumental and formulaic and damaging to certain factions of urban life. The Creative City (with a capital 'C') therefore will be used as shorthand for the capitalistic, paradigmatic (bordering on dogmatic) and meta-narrative view of how creativity can be used to economically stimulate and develop the city (an argument developed further in Part I). In reaction to this, the book will champion a different kind of city (in Part II). This is not an attempt to go 'beyond' the Creative City by furthering its economic determinism, such as arguing for the use of cognitive-cultural capitalism instead (Scott 2014). Such an argument, however valid from an economic geography perspective, only serves to replicate the existing problem of the Creative City. Nor is it an attempt to replace an economistic view of creativity with one that focuses on a creative ecology of public-private initiatives, social engineering and place-making (Landry 2011). Such a view broadens the actors involved but creates similar outcomes. Nor it is an attempt to argue that the Creative City should re-engage with the notion of culture and publics (Vickery 2015). Rather, this book will champion the creative city (with a small 'c'). Such a city embraces experimental and creative social interaction, and broadens our usage of urban functions, not constricts them, producing more heterogeneous and less homogeneous urban spaces – a city where skateboarders who reappropriate the 'dead space' of the undercroft area of the South Bank in London are not removed because the landlord wants to build more coffee shops, restaurants and other homogenous retail outlets; a city where squatters who have occupied a derelict area of Copenhagen and fostered an artistic community that serves local residents are not the focus of an eviction campaign; a city where recreational trespassers who are caught on top of a New York skyscraper are not arrested and charged; a city where the removal of inappropriate corporate advertising material by local residents is not criminalised; a city where if people want to shoot a short film on an under-used privately-owned plaza do not need to submit written notification and wait three weeks for a reply; a city where being creative is the very act of citizenship.

Methodologically, the book uses data gathered from over six years of travelling to cities around the world. Primarily, the book will draw on visual qualitative data from over 25 cities in ten countries on four continents (a collection of over 3,000 images and 200 videos), but also from a plethora of items gathered including promotional pamphlets, community papers, glossy magazines, discarded artistic artefacts and essentially anything that alludes to the creative practice of urbanites (official and illicit). Formal and informal interviews were carried out with people who instigate Creative City policies such as government officials, private companies and creative industry workers, as well as subcultural practitioners, cultural consumers and activists. Enlivened by vignettes, stories and explorations, this book weaves these together into a comprehensive 'lived' ethnography of urban experiences. In a 'flânerie' way, I encountered creative industry practitioners, policy-makers, academics, urban managers, architects, planners, subcultural participants and generally creative citizens, and registered (visually, textually and emotionally) their activities and engagements as they were performed across various spaces and places in many city that have been visited and lived in (see Kramer and Short (2011) for a more detailed explanation of this urban methodological practice). In some cases as an observer and in others as a fully fledged participant, I have witnessed first hand acts of urban creativity (subcultural, subversive, capitalistic and otherwise), experienced directly the Creative City paradigm in action (and in some cases contributed to it) and personally explored the urban environment that it effects. For example, I conducted parkour (irregularly) for a period of nine months in 2008, I have undertaken a number of urban explorations with practised participants and been involved heavily in campaigns to save subcultural spaces from 'development'. Conversely, I have also worked for policy units that aimed to bring about certain factions of Creative City policy; I did internships in creative economy firms in Sydney for a period in 2005, and have consulted for specific municipal governments of creative strategies (namely Sydney, Tel Aviv and London). Combining these experiences makes for a holistic view of creative practices, both those that have contributed to economic development as well as those that try to resist it. All these data then will be used to exemplify the arguments put forward throughout the book and colour the debate with contemporary case studies relevant to the theory being discussed. While using vignettes from cities such as Seoul, Leicester, Shanghai, Bristol and others, there are a few key cities that stand out for special scrutiny. Given my intense study of Sydney's Creative City policies and my interaction with them (as well as the amount of information that it produces), it was a suitable city to linger on. Also, London is used overtly as a city of study because of the amount of time spent there, but also because it is a city at the 'forefront' of economic globalisation (and all the problems therein entailed). For similar reasons, New York is used frequently. Also, given my work, experience and situ in the United Kingdom's national political landscape, many of the UK's Creativity-related policies will be analysed and utilised to further the arguments of the book.

Such geographical eclecticism could lead to accusations of empirical superficiality; however, it must be stressed that this book is a theoretical exploration of what the Creative City is, how it purports to help but ends up hindering and homogenising creative practices, and how we can think about subverting it. It touches upon empirical examples to illustrate particular points, but the empirics do not drive the argument. Moreover, given the nature of Creative City development and the fleetingness of urban subversion activity, any empirical observations will be rendered irrelevant within the course time. The Creative City paradigm is constantly changing its characteristics, and with it, so too is urban subversion changing as it reacts to it. Therefore to linger too much on empiricism would be to obfuscate the theoretical analytics of these ever-increasingly rapid changes. So, this book is meant as a way of navigating the inherent problems of the Creative City and to theoretically explore the way in which a creative city can be sought.

Moreover, there are a number of theorists, philosophers and key urban thinkers that are alluded to in this book. But again, I have utilised specific ideas from them in order to forge a course toward the realisation of a creative city. Granted, some ideas have greater influence than others (such as Deleuze and Guattari, for example), but it is because these ideas lend themselves most appropriately to the way in which the argument of the book progresses. As a result, the book is not aiming at questioning the politics and full theoretical *oeuvre* of any one (or more) author, such endeavours are more fully conducted elsewhere. Instead, specific ideas, theories and philosophies are utilised first and foremost to drive my argument. Moreover, the book should not be read as some sort of 'Deleuzian' or 'De Certeauian' account of urban creativity, but nor should it be approached as one without a theoretical mandate. On the contrary, the argument of this book forges its own theoretical disposition that draws from the crowd, and one that would see all urbanites members of a creative city.

The book is therefore structured into two main parts, which together chart the conceptualisation, critique and potential futures of the Creative City. Part I is a thorough archaeology of the Creative City paradigm and its incumbent ideologies and ontologies. It charts the growth in popularity of the idea and outlines the key protagonists of the paradigm and how it became a formalised and standardised 'model' that could be replicated across the world. To do so, though, it is necessary to delineate the concept into its three main (but of course, interrelated) constituent parts. Chapter 2 is a history of how the Creative City concept and the associated vernaculars were foregrounded. More than simply a novel idea that was conceived in isolation (as is often the perception purveyed in the critical literature), the Creative City was as *systemic requirement* of contemporary capitalistic urbanism. As outlined by Harvey (1989) city government structures have come to resemble corporate frameworks rather than postwar urban management structures. The outsourcing of specific services to private companies, the increasing erosion of public spending in lieu of private financing and the use of branding and marketing 'gurus' are all in aid of making a city more competitive on the global scale. This is linked to the increasing preponderance of neoliberalism as part of contemporary urban agendas, which has seen it become a key philosophy of how cities attempt to manage and develop themselves (Brenner and Theodore 2002). Relatedly, the 'Global City' idea, popularised by Sassen (2001) and then put forward methodologically through the creation of a hierarchy of cities by Taylor (2004), has created a 'race to the top' where cities compete with each other for the global flows of capital and finance. The 'list-mania' that pervades contemporary urban policy now fuels the need for the next competitive edge and valorises the continual search for the 'new'. Cities, because of this neoliberalisation of urban governance, are now in competition with each other. There is hence a continual search for new ideas, novel ways of attracting capital and extracting the surplus value needed to obtain that critical advantage. Therefore the contemporaneous preponderance of the creative industries as a political economic tool was seized upon, creating a 'perfect fit' for urban governments across the world. Hence, an idea as neatly packaged and easily employable as the Creative City is devised as the spatial manifestation of a number of prevailing political vernaculars. So far from simply being formulated in response to the observable phenomenon of an increase in the creativity of urbanites, the Creative City idea was *required* in order for cities to maintain an economic competitiveness and hence the theories and ideas were moulded to suit.

If the Creative City is the Trojan horse, then the creative class are the Greeks inside. Chapter 3 outlines how Richard Florida's theory of the creative class came to dominate urban policy in the early part of the twenty-first century. The chapter outlines the evolution of the idea from the economic geographical and regional development origins, and how when applied practically it has severe limitations. There are of course various other critiques of this work, which are equally as important in showing how the creative class thesis is related to and indicative of the Creative City as a tool for property and real estate development. The critiques are multifaceted, ranging from the economically based critiques on the oversimplification of urban creativity as an economic growth strategy, methodological issues surrounding the conclusions inferred from the data, and more theoretical misgivings as to the way in which the ideas have been unreflexively taken up by a multitude of urban, local, regional and national governments around the world. However, more recently, the creative class thesis has been a rubric under which cities have gentrified certain areas, and 'upscaled' urban locales that are seen to be in need of it. This 'creative classi*fication*' of cities has been the subject of intense debate, particularly online and on blogs, which is from where Florida now transmits the creative class ideologies. Moreover, this chapter asserts that while the creative class theory has been thought of as simply a response to a new configuration of economic behaviour independent from the ideas outlined in Chapter 2, it is in fact again a systemic part of the Creative City paradigm.

The urban development that is designed with the creative class in mind is the focus of Chapter 4. If the numerous Creative City branding exercises are the 'soft' policies, then the 'zoning' of creative practices in the city are the 'hard' infrastructural policies. And it is this part of the Creative City paradigm where the most value and profit is to be extracted. Because of the prevailing rhetoric, these physical manifestations of the Creative City are specifically designed to attract the creative class, and as such, have an associated language and narrative that neutralises the impacts that such developments have on the existing urban locale. The issues of gentrification will be developed in order to have a wider appreciation of the ways in which Creative City-led developments have deleterious effects. Through a discussion of Cultural Quarters and Media Cities, this chapter highlights how and why the zoning or 'quarterisation' (and forced agglomeration) of creative and cultural activities is part of the 'serial replication' of the Creative City rhetoric and detrimental to other (non-economic) creative practices, urban subversions and the wider social and community fabric that exists in the area being targeted. These developments often have a linguistic veneer of encouraging creativity and fostering the all-important 'creative atmosphere' that is the perceived pull factor for the creative class (and the economic benefits they supposedly bring), yet they are driven by financial resources and a political economic narrative of real estate development and wealth generation. As a consequence, they actually negate such a creative 'buzz' that the narrative claims to inculcate. This feeds into the wider process of urbanisation that Lefebvre (2003 [1970]) outlined, and as such it is shown that the Creative City paradigm cannot be unhinged from the urban entrepreneurial process, despite evangelistic claims to the contrary.

Part I concludes by highlighting how the Creative City paradigm, as it has been described (i.e. economically deterministic), fails to include any causal mechanisms of creativity at all. In other words, the Creative City paradigm arose out of a need for a global competitiveness which included an identifiable image and holistic city-scale validation of infrastructural projects, and not because it was a characteristic of the collective creativity of urbanites. As such, the Creative City concept is flawed and through a critique of its systemic construction now lies in metaphorical ruins.

If Part I of the book has deconstructed the nature of the Creative City and argued that it has no causal mechanism for creativity, then Part II of the book argues that we can 'rescue' the idea of creativity. This, however, requires a radical shift in the definition of what it is to be creative; it requires a focus on urban creativity beyond that which is prescribed by the 'Creative City' narrative. So Part II attempts to redraw a creative city, one that is ontologically, epistemologically and empirically different to the Creative City. Chapter 5 couches this argument through a history of urban interventionism and a discussion of some illuminating ideas. By revisiting early sociological accounts of urban recreation and resistance to oppressive urban development, this chapter (mirroring Chapter 2) 'sets a scene' from which a creative city can be realised. It provides not only the conceptual touchstones from which we can envision a more creatively equitable city, but it also gives us a language that makes it more challenging to replicate the problems of the Creative City. Moreover, the chapter discusses flâneurism, the Situationists and unitary urbanism, and later DeCerteau (1984), to argue that creativity has always played a part in everyday urbanism, if not expressly idealised as such.

Chapter 6 explains how the term urban subversion can be used to encapsulate those activities that could be seen to have their ontological influences in the historical discourses outlined in Chapter 5. The term urban subversion is a contested one, often used as a pseudonym for creative urban subcultures. While indeed there are some creative subcultures that embellish the properties of urban subversion, they are a more identifiable and therefore 'risky' articulation of urban creativity, 'risky' in that once a subculture can be identified, it can be 'recaptured' by the Creative City. So, through a detailed analysis of urban exploration and street art (and touching upon other subcultures), this chapter analyses how urban subversion is a multifaceted process of continual critique of not only the Creative City but also other problematic institutions that can be created in response to it. It explains the way in which urban subversion can be thought of as the initial 'spark' of creativity that realises an alternative function to the city, but also how this can be 'subculturised'. It will detail that the term urban subversion is less a proper noun and more a 'state of mind' that requires a constant questioning of the established systemic order of the Creative City and indeed any other structural inequalities that are created (including within a subculture). Using social theoretical arguments based on Deleuze, Baudrillard and others, the chapter argues that *true* creativity shown by subcultural participants comes about because of the ways in which it is not aligned to a wider system of functionalities. In other words, being truly creative (and realising a creative city) requires engaging in urban subversion at multiple times, locations and conceptual levels.

Chapter 7 outlines how urban subversion has an uneasy relationship with place. The ossification of subcultural activity requires the identification of a location, a place to write histories and stories to claim as their own. However, by doing so, they render themselves more susceptible to the marginalisation forces of the city. More damagingly perhaps, it also renders them more liable to be co-opted by the Creative City as the latest iteration of novelty to be profited from. Many subcultural places (or 'spots') have been created, with very different reasons, mechanisms and outcomes. By focusing on London's 'sliver of subversive subcultural spaces', this chapter focuses on the how urban subversion conflictingly relates to the physicality of the city. In addition, the city has begun to house creative subcultural activities in purpose-built 'zones', further developing a reactionary ethos within those urban subversions not wanting to be compartmentalised. These institutionalised places represent the way in which the Creative City can manifest subversion for its own development agenda.

The book's final chapter is then an exploration of what kind of cities might materialise as a result of more or less urban subversion, and its different interpretations. Using existing theoretical tropes from Boltaniski and Chiapello, Lefebvre, Mouffe, Baudriallard, Deleuze and Guatarri, and Badiou, this chapter explores the theoretical implications and limitations of urban subversion. Far from being predictions or a manifesto for the creative city, I attempt to highlight the different and varied ways in which the juxtaposition between urban subversions and the Creative City can be viewed, theorised and envisioned. The creative industrial city, the spectacularised city, the creatively activist city and the socially creative city will be detailed and discussed, with empirical examples again alluding to how they might not be as ontologically distant as first thought. The chapter also concludes the book with thoughts about further avenues of research and analysis.

This book is therefore a theoretical exploration of what it means to be truly creative in the city. The approach is holistic, in other words I have preferred abstraction, subjectification and theoretical postulation rather than a reliance on specificity, intense empiricism and objectification. This approach of course has its limitations, but it is most appropriate given how preponderant the Creative City paradigm has become in recent years. Put crudely, to deconstruct the gargantuan behemoth that the Creative City has become requires thinking big. So, let us begin.

Part I

2 Creating a scene

The need for a Creative City

Almost exactly one year after I was sat in that café on King Street in Newtown, having been asking 'what creative life do you want for Sydney?' (see Figure 1.1) for that 12-month period, the City of Sydney released a document called 'Creative City: Draft Cultural Policy and Action Plan 2014–2024'. The document (which is available online) covers the cultural policies that the city plans to implement over the next ten years. Using the language of culture and creativity throughout, the document is the city government's effort to identify the way in which Sydney can develop its cultural provisioning. It states at the outset that:

> Culture is widely recognised as a major part of Sydney's success and global city status. We are geared to finding new ways for Sydney to build its creative muscle so that creativity and cultural connections are the essential tools to meet the challenges of urban living. (City of Sydney 2014: 8)

The plan is the culmination of a lengthy consultation period, which was based upon the April 2013 release a 'Creative City discussion paper', in which 'more than 2,000 voices' (City of Sydney 2014: 8) shared their views on what was needed for Sydney to 'unlock the creative potential of its city and its creative community' (ibid.). The document reveals that the outcome of this consultation process was a valorisation of the city's existing creative and cultural initiatives (broadly defined), as well as some new 'wants' for the city. For example, the document articulates that what was called for by the consultation process was for 'creativity to be more frequent and visible in the city's public domain and its precincts through a critical mass of activity' (ibid.: 9). In response to this demand, the city will support the development of 'cultural precincts' and continue support for the 'Cultural Ribbon' project. This project links together existing cultural institutions that are geographically proximate with the Darling Harbour area of Sydney. The idea of a Cultural Ribbon will be a 'walking trail and more' and it will offer a 'cool way to explore the city' (City of Sydney n.d.: n.p.). The idea of a Cultural Ribbon therefore seems to be an attempt to tie together already existing cultural and creative infrastructures through a new marketing and branding initiative and to encourage people (the wording used in the document is 'visitors' and 'enhance the visitor experience' (City of Sydney 2014: 39)) to view such offerings in new participative ways. Also, other cultural precincts will be nurtured through tackling key issues such as 'connectivity to the city and accessible transport solutions' and the 'integration of appropriate business/retail mix including night-time/daytime offerings' (ibid.: 39). Such wants are seen as deliverable through the 'Actions' section of the document, which outlines specific 'new' activities such as to 'develop the city's urban planning function and capability to plan or influence cultural precincts and infrastructure' (ibid.: 43). It also speaks to the utility of existing projects such as 'culture-led revitalisation of the Oxford Street precinct including curated creative retailing [and] affordable creative spaces' (ibid.: 44).

Also in response to the apparent desire of Sydney's residents for more 'visible creativity', the city will:

> Initiate creative projects with business and other city stakeholders to bring creativity into the everyday experience of the city (such as artwork commissions on building site hoardings), and by animating public spaces with simple ideas (such as deckchairs or hammocks in new parks and plazas or table tennis tables in urban spaces). (City of Sydney 2014: 10)

Hence, by instigating ideas that 'animate' the public sphere, such as deckchairs, hammocks and table tennis tables, the City of Sydney hopes to encourage people to engage with these spaces more creatively and culturally. The plan details other initiatives that the city council hopes will encourage more creativity throughout the city including continued funding for cultural events, more 'affordable' spaces for artists and educational strategies for school children.

The entire document, all 118 pages, is riddled with a language that engenders a city that is on the cutting edge of 'creative' practice. It says it has listened to Sydney's residents by asking them 'what creative life do you want for Sydney?' (see Figure 1.1), and is now embarking upon ambitious new initiatives that will inspire a generation of Sydneysiders to be more creative – thereby engendering a 'liveable' city. It will formulate 'cool' ideas like hammocks in public spaces, create cultural walks and tinker with policies to release creative economic practice from the current shackles of an overly bureaucratic administrative process. It uses phrases like 'creative retailing' and 'business/retail mixes' to articulate an assumed novel form of economic activity, and links previous infrastructure projects (that have existing social impacts) as part of the overall strategy to make the city creative. Overall, it reads in a way that suggests that creativity will solve the city's issues, and forge a new path to a more prosperous and vitalised city. Landry and Bianchini penned the Creative City manifesto in 1995, suggesting that 'creativity can be mobilised to help solve the myriad problems of the city' (Landry and Bianchini 1995: 9), and nearly two decades later, Sydney is doing exactly that. But the 'creativity' that it is mobilising is a pastiche, a linguistic simulacrum that is empty of any subjective meaning of 'creativity' and represents nothing more than the political and economic vernacular it has become. A more critical look at the document shows that many of the initiatives are focused in the city's CBD, speaks in the vast majority to artistic economic endeavours and has little to no mention of the social problems inherent in the city's minority and indigenous populations. The definition of 'culture' within the documentation and the wider political mandate is somewhat limited to that which is conducive to the economic urbanisation process. There is no mention of minority cultures, subcultural activities or the large variance in Sydney's ethnicity. It advocates existing projects (like the Cultural Ribbon) by arguing it is what was 'called for' in the consultation process, and partakes in a vast swathe of 'place-making' activities that beautify public space, but is short on actual policies that aim to address the root causes of the current lacuna in cultural participation. It redresses existing projects as 'creative' in order to obtain more financial backing, and repackages more 'mundane' activities such as retail, condominium and 'mixed-use' business unit construction as creative, cool and innovative. As such, the document is a distinguished example of a Creative City policy. It highlights the ways in which creativity has been used as a linguistic tool to valorise existing infrastructural projects and justify policies of beautify-ing public space, real estate development and financialisation of previously non-business cultural activity.

This kind of document is of course, not limited to Sydney. Similar documents have recently been compiled for countless other cities around the world, some perhaps to be expected, such as Vancouver, Toronto, Baltimore, Seoul, Milwaukee, Copenhagen, Shanghai, Jakarta, Sheffield, but also perhaps some cities that one would not expect, such as Kanazawa, Tehran, Flanders and Flint. And these are just the examples that are freely available online and are written in English. There is a plethora of policy literature that all purports to similar ideals – to implement policies of 'creativity' to help upscale the city's social well-being, infrastructure, cultural participation and economic vitality. However, like Sydney's Creative City Plan, they also use the language of creativity uncritically, using it as a byword for positivity and unproblematic functionality. Such language has become 'fast urban policy' (Peck 2005). This means that they are the policies of contemporary urban governments that are designed less to tackle the root causes of many difficult social, cultural and economic challenges that they face via cultural and creative means, and more to excuse and justify activi-ties that promote and valorise economic production and profit-making for inter-ested private (and public) stakeholders at the expense of everything else. And it is on such documentation that the *Creative City paradigm* is predicated. It is a *paradigm* because it has been espoused by an amalgamation of a suite of processes that fall beyond the purview of any one institution but are replicable, patterned and predictable. It is no longer the remit of a government (urban, regional, national or any other scale that can be attributed), but of governance (Jessop 1997). This means that there is a range of actors and institutions from the public, private and community spheres that are all collaborating (as well as competing) to not only manage, maintain and control the urban, but to *produce* it. Edensor *et al.* (2009) have rightly asserted that the city is the sole arena for such creativity agendas. They state that, 'spaces beyond the urban or non-metropolitan [are] ignored or trivialised' (ibid.: 5) by these policies. But such a focus on the city is a fundamental prerequisite of the creativity agenda because it is part and parcel of the wider urbanisation process.

This notion will be excavated intricately over the next few chapters but initially, a fundamental notion that is critical to the realisation of the Creative City paradigm and its governance is the shift from an urban managerialism to an urban entrepreneurialism. This 'shift' toward urban entrepreneurial activity was noted by Harvey (1989) when he argued that 'the managerial approach so typical of the 1960s had steadily given way to initiatory and entrepreneurial forms of action in the 1970s and 1980s' (Harvey 1989: 4). Such approaches, he noted, were characterised by economic growth strategies, promotional campaigns and the support of local small businesses that were necessitated by a reduction in funding from national or federal pots that were themselves being 'hollowed out' through Thatcherite policies and the onset of 'Reaganomics' (see Frank 1982). The process of urban entrepreneurialism then is linked to the wider ideology of 'neoliberalism'. Therefore to begin our excavation of the Creative City paradigm and its inherent problematic characteristics, it becomes necessary to understand neoliberalism within a broader context. Also, given that the term has a seemingly varied understanding within the urban governance literature (Springer 2010; Davies 2014), it is worth unpacking the concept in the next section.

Neoliberalism

The Second World War left many Western countries battle-weary. The loss of large populations of the workforce, the destruction of urban and industrial infrastructure and the general fatigue of six years of the horrors of war had a profound social, but also massively deleterious economic impact. So in 1944, the 44 allied nations agreed to formulate the Bretton Woods institutions (the IMF, the World Bank, etc.) in the hope that money for reconstruction projects and the need to smooth out balance of payment deficiencies would negate the need for any future global conflicts. Concurrently, the Marshall Plan created a strong trade link between Europe and the US, and many countries in the then-Third World were becoming decolonised. All these processes were giving way to immense financial and economic opportunities. No one saw this more than Friedrich von Hayek who was based at the University of Chicago. The ideas of Hayek were concerned with reigniting liberal politics of minimal state intervention (following the philosophies of *liberalism* of John Locke), after welfare state policies had suffered perceived crises. Therefore, engendering a *neo*liberalism, Hayek's economic philosophy of free market competition and a minimal state was in stark contrast to the current prevailing political ideologies of a strong welfare system and a 'big' government in the postwar period (put forward by John Maynard Keynes). Through gathering a support network of key politicians from around the world, influential commentators and thinkers, Hayek's ideas slowly began to percolate through to mainstream political thinking (see Klein 2008). These ideas soon became the basis for entire national-level political agendas, and when one of Hayek's most keen observers and admirers, Margaret Thatcher, became British

Prime Minister in 1979, the UK began a transition to a neoliberal agenda. This was mirrored in the US when another one of Hayek's admirers, Ronald Reagan, became the 40th President of the United States in 1981.

The ideological notion of neoliberalism can be attributed to a set of economic philosophies that Harvey (2007: 22) suggests are 'proposing that human well-being can best be advanced by the maximisation of entrepreneurial freedoms within an institutional framework characterised by private property rights, individual liberty, unencumbered markets, and free trade'. Decrying neoliberalism as a 'class project' (Harvey 2009: n.p.), it is about recasting individual behaviour toward self-fulfilment and the emancipation of the 'enterprising self'. It is predicated upon an innate preponderance of competition, which pervades all forms of societal exchange (Davies 2014). Other analyses of neoliberalism emphasise the way in which it naturalises market relations (Peck 2001). Accounts of neoliberal strategies in relation to the Creative City have articulated how austerity measures centralise wealth and appropriate the resultant marginalised poor (in squatter settlements for example) as 'cool' and 'edgy' parts of the city (Mayer 2013). There is a universe of literature that could be cited and explored when unpacking neoliberalism, but suffice it to say it is multiplicitious, complex and variegated. To think otherwise is to reduce neoliberalism to an external monolithic and Leviathanical force which only serves to create mirror images in countering it. Such a singular and overgeneralised notion of neoliberalism has been rebuked, as such a conceptualisation is inadequate in catering for the proliferation of local variegations, nuances and subtleties that at any one time compose the neoliberal project. It is more important to recognise how the process of *neoliberalisation* is mobile across the world agnostic of national, regional or urban boundaries, but also how it gains particular 'tractions' as it travels, and takes on variegated forms as it coagulates with existing conditions 'on the ground' (Ong 2006; Springer 2010).

Despite these conceptual and geographically reflexive nuances, neoliberalism is often used as shorthand to denote the overall ethos of market-led strategies, the formulation of individualisation and economic self-sufficiency as core truths, reduction of welfare to as low levels as possible, active resistance of unionisation and the forcing of the marginalised (i.e. those that cannot contribute to the rapid accumulation of wealth) into low-wage, unprotected employment (Peck and Tickell 2002). It has also been accused of having a distinctly masculine bias, and that the neoliberal development has marginalised women, ethnic minorities, non-heterosexuals and disabled individuals (Hubbard 2004). Such critiques are valid accusations to level at what is a highly elitist process of development. But this shorthand should not be mistaken for the simplification and non-recognition of the multiplicitous variance that neoliberalism entails.

Neoliberalism, as a suite of political and economic processes, has of course had a profound impact upon the form and functioning of cities. Lefebvre (2003 [1970]) notes how the transition toward an industrial city in modern times has been characterised by the homogenisation of cityscapes and the serial reproduction of the built form – in other words a city built by industrial processes (homogenised on the factory line à la Fordist production) rather than a canonical city full of industry. A fundamental catalyst of the industrial city has been neoliberalism, and continues to be so:

> Across the global urban system, the widespread adoption of all-too-familiar neoliberal discourses and policy formulations is connected to a more deeply rooted and creatively destructive process of diachronic transformation – of policies, institutions and spaces – that is mutating the landscapes of both urban development and urban governance. (Peck *et al.* 2013: 1092)

This 'mutation' referred to here has very particular characteristics, the most obvious of which is the privatisation of public space via urban governments auctioning off land to private real estate investors to increase funds (Smith 1996). This core process of neoliberal urban strategy has led to auxiliary consequences of increased securitisation of urban space (as private companies look to protect their investments (see Graham (2009)) and the active marginalisation of the poor (Watson 2009). So, as private, multinational real estate companies increase their role in the formulation of cities, urban space becomes another commodity in which profit can be made. The reduction of costs by the replication of design (i.e. copy-and-paste production) becomes important, as does the increase in price through a reduction in state-led rental controls. The economistic, instrumental and financially deterministic process of corporate vernacular becomes intertwined with urban space and its social and cultural idiosyncrasies. As such, current urban growth strategies, pregnant as they are with neoliberal philosophies, create urban spaces that are designed first and foremost to produce profit and wealth. They also create cities that cater most readily to white, middle-class, heterosexual, able-bodied men. The main cohort of people who design, manage and maintain cities are drawn from the elitist power networks that characterise modern-day Western governments, and as such the urbanisation process reflects their character. It mirrors and reproduces the hierarchical patriarchies (and the systemic inequalities that such a hierarchy espouses) that have come to characterise late capitalist, liberal democratic societies. Neoliberal philosophy is predicated upon the exploitation of the marginalised to the benefit of those in the 'centre', and as such there is gendered, racialised, heteronormative bias to neoliberal development in the physical, emotional and social design of both private and public spaces in cities. In addition, such spaces are characterised by conspicuous consumption (Harvey 1989), high investment levels and rapid rent increase (Lees 2003a), the degradation of public, civic and community activity, and the securitisation and decreased access to public urban space.

  Neoliberal policies mutate, transform and shift over time and space; they are what could be called 'assemblages' that have an international mobility (McCann and Ward 2011). The idea of an assemblage, as MacFarlane (2011: 653) has noted, 'is increasingly used to connote, expansively, indeterminacy, emergence, becoming, processuality, turbulence, and the sociomateriality of phenomena'. Assemblage theories are drawn from a Deleuzian ontology of *becoming* which focuses more on the processual and emergent qualities of matter, rather than a more solid state of *being*. Such a complex ontology of *becoming* will be alluded to more in Part II, but suffice it to say that assemblages and assemblage theory have become more prevalent in recent urban studies literature as a means of articulating the fluidity, turbulent emergence and non-linearity of urban phenomena. Characterising neoliberalism as an assemblage then is a deliberate attempt to transcend a linear thinking that traditionally sees neoliberalism and its inherent political processes as a cohesive, globalised and singular whole – a Leviathan to be countered. More sophisticated accounts have articulated the adaptive nature of neoliberal policy, its international mobility and ability to evade political and/or social co-option. As such, neoliberalism has adapted to the changing landscape of urbanity, including resistance and subversion to it. Urban managers and elites are constantly searching for the 'new', the novel policy or innovative urban development gimmick that could be transmogrified into a new locality (with maximum efficiency) to produce further wealth. By constantly co-constituting and amalgamating disparate policies into new places but with subtly different superficial characteristics, the urban governance structures (i.e. the public-private nexus of institutional powers that urban 'government' has become) are creating urban spaces that become policy playgrounds for experimental neoliberal agendas. It is this flexible adaption that has rendered neoliberalisation such an effective and above all profitable mechanism for urban development.

As will be noted throughout this book, while the *language* of urban neoliberal politics is ever changing, the fundamental philosophies remain more resolute. Peck (2005), perhaps the most vehement and (in)famous of all critics of the creativity rhetoric, has outlined how it is far from a novel and innovative means of conducting urban governance, but simply the latest iteration of prevailing neoliberal urban development. The language and political articulation of creativity has become critical to urban governments across the world, but it is part of a much broader narrative that includes neoliberalism. The concept of the 'Global City' is also important to the contextualisation of urban development and has itself become susceptible to these 'changes' in language. It is again highly compatible with and embroiled within the Creative City paradigm and so requires a closer inspection.

Global Cities

The old classical saying is that 'all roads lead to Rome'. The economic, cultural and political centre of the Roman Empire, Rome, like Athens to Greek civilisation, was the beacon of modern life, a place of civility, progression and enlightenment to which the rest of the world was to aspire. In our neo-colonial, hyper-connected world of the twenty-first century, the desire of cities to have a similar global dominance has not waned. There is, and perhaps always has been, an innate need for urban centres to be considered 'Global Cities', to be viewed as key sites in the progression of modernity and centres of economic might, cultural dissemination and political power. And so, in contemporary urban politics, we have the term 'Global City', a label that has been brought about through academic and political rhetoric, and a title which cities are desperate to obtain.

Take Tel Aviv, for example. An Isreali city of around 400,000 people on the Eastern shores of the Mediterranean Sea, Tel Aviv is a cosmopolitan hub, known as a liberal city in an otherwise religious conservative region. In 2010, the mayor Ron Huldai started the Tel Aviv Global City initiative, which is a 'national initiative aimed at elevating the city's global positioning' (Tel Aviv Mayor's Office 2010: n.p.). Run by a small team of municipality officials, the Tel Aviv Global City unit is charged with extolling the virtues of Tel Aviv to an international audience, to attract businesses and tourists, and generally market the city globally. The skillset of the team ranged from marketing to public relations, and through visiting international conferences, inviting key thinkers and urbanist academics and engaging with media outlets, the small team have made a large impact upon the international community since its formulation in 2010. The unit has put together promotional material that showcases Tel Aviv's business strengths, nightlife, tourist facilities, cultural and religious heritage, young demographic, liberal and cosmopolitan attitudes, educational and academic excellence and its entrepreneurial capacity. The promotional material online and in print highlights the recently built new building for the Tel Aviv Museum of Art, the Design Museum in Holon, its vibrant and tolerant nightlife and upscale tourism facilities (including hotels, business centres, taxis and so on), the UNESCO heritage site of the White City, the city's beaches, its start-up business culture, tax incentives for foreign investment and other cultural facilities and institutions.

The Tel Aviv Global City unit has not built anything directly, nor does it offer tax relief for foreign investment. It has not serviced Tel Aviv's physical infrastructure, nor has it offered any educational, social or public services. Yet it has done more for the visibility of Tel Aviv within policy and international business circles than could have potentially been achieved via more 'traditional' mayoral and urban government methods. In effect, Tel Aviv Global City is the marketing department of the city. Its remit is to 'elevat[e] the city's global positioning' (Tel Aviv Mayor's Office 2010: n.p.) and it is achieving this with remarkable fortitude in the urban policy arena, particularly with regard to the city's innovation and start-up capacity. It has been recognised as the second best city for start ups behind Silicon Valley in the US, and came second in the Innovation City award organised by the Urban Land Institute. Through the Tel Aviv Global City unit, the city is marketed and promoted in a similar vein to large international corporations. It is a prime and very visible example of urban entrepreneurialism, where cities act increasingly business-like in a globalising, competitive world.

The term Global City has a very complicated etymological and conceptual history. It is theoretically linked to the term 'world city', which was first coined in its current incarnation by Patrick Geddes in 1915 in his book *Cities in Evolution*. He uses the term lightly, however, and it is not defined beyond the amorphous identification of an urban cosmopolitanism that he witnessed in his travels to London, Paris, New York and a few other major metropolises of the time. Over half a century later, the late Peter Hall wrote in 1966 *The World Cities* and offered a far more distinctive definition of what a *World* (not a Global) City entails. In the book he argued that World Cities are 'certain great cities, in which a quite disproportionate part of the world's most important *business* is conducted' (Hall 1966: 7, my emphasis). We begin to see here the economisation of the term, with business being considered the important variable in making a city 'worldly'. Hall subsequently went on to advance these ideas to suggest that World Cities are centres for consumption, the arts, cultural diversity and political power, and cited London, Paris, Randstad Holland, Rhine-Ruhr, Moscow, New York and Tokyo as the seven 'World Cities'. This neo-Marxist view of a city's international appeal and reach being predicated upon an underlying political economy is one that has dominated urban theory to date.

While the World City term became a popular label for cities wanting to market themselves as international 'centres' the idea of the 'Global City' was to be even more so. Ever since Saskia Sassen wrote a book called *Global City* in 1991 (revised in 2001), the term has infiltrated much of social science and humanities literature and has become a buzzword in urban policy realms, primarily because of its etymological link with (economic) globalisation, which cities are competing to be part of and to benefit from. The idea of the Global City was qualitatively different to that of the World City in that Sassen (2001), building upon ideas purported by Hymer (1972) and Friedmann (1986), argues that Global Cities are 'global' because they *command and control* the world economy. If a World City is a collection of 'worldly' institutions such as international banks, museums, art festivals and highly cosmopolitan and diverse cultural and ethnic populations, then a Global City is far more instrumental and systemic to the process of globalisation. It houses specific financial institutions that have a direct effect on what happens throughout the rest of the world's national, regional and urban economies. The term 'Global City' has been adapted and hollowed out somewhat by mainstream media narratives, but Sassen's original conceptualisation of a Global City was one of financial and economic control. She argued that;

> ... these cities now function in four new ways: first, as highly concentrated command points in the organisation of the world economy; second, as key locations for finance and for specialised service firms ...; third, as sites of production ... ; and fourth, as markets for the products and innovations produced. (Sassen 2001: 3)

These four functions, wedded by a conceptual cornerstone of agglomeration of the headquarters of advanced producer service firms (shortened over the years to APS firms), certain cities (she identified London, New York and Tokyo as the triumvirate of contemporary Global Cities that represent the three key global regions of Europe, the US and Asia) have an overly strong influence on the flow of finance, goods and labour throughout the world. Many APS firms have headquarters or high-level offices residing in these Global Cities. As such, the key decisions about the location of global capital (financial and human) are made in these locales (Taylor *et al.* 2010).

The work of the Globalisation and World Cities group (GaWC) at Loughborough University directed by Peter Taylor put forward Sassen's conceptualisation of the Global City (although in doing so inversed the theoretical definition to one whereby the World City Network (Taylor 2004) was the basis of a city's power, not the agglomerative potential (see Smith 2014)). Via data-intensive and calculative methodological research on the office locations of the world's largest APS firms, GaWC devised a ranking of cities based on their 'networked-ness' within the world economy. The delineation of particular cities into 'alpha', 'beta' and 'gamma' cities (the classification chosen to mirror that used for celestial stars, as it was a 'universe of cities') was a particularly important methodological intervention, as it gave a quantifiably empirical justification to the hierarchies that already existed, albeit rather arbitrarily, with Hall (1966) and Sassen (2001). Earlier quantitative rankings of cities from economists included the 'hedonic' approach (Rosen 1979) and the 'revealed preference' approach, which rather bafflingly 'can rank city quality of life even if no attributes of the city are observed' (Kahn 1995: 224). But in terms of producing hierarchies of cities in an era of globalisation that were based on actual observed phenomena, GaWC had brought a more nuanced and empirically accurate approach to city rankings, and as a result city governments around the world began to take merit in being highly positioned in the rankings (see also Godfrey and Zhou 1999). The economistic view of a city's status in the world and the growing propensity for ranking cities has fuelled demand for different types of city league tables.

Such quantification is critical in valorising urban hierarchies as it gives an empirical and codified credence to such lists, an analytical, quantitative and logical reasoning that removes subjectivity from urban practices. As such, the ideology of the Global City is one predicated upon the presupposition of financial and economic might. The very existence of the concept (and the associated methodologies and epistemologies) is shorthand for being a city that *matters* for economic globalisation (and also an instigator of global income inequalities), and such presumptions are better founded upon seemingly logical quantitative empirics, i.e. rankings. The Global City mantra hence is founded upon the 'tyranny' of rankings (or list-mania), and creates a zero-sum game that allies so affirmatively with contemporary entrepreneurial and competitive urban politics (Harvey 1989). It is a language therefore that has gained traction with urban governments. The term soon infiltrated public urban policy and over a short period of time, city governments all over the world became fixated on where they were placed in the rankings and what they needed to do in order to climb them (Giffinger *et al.* 2011) and attain the status of Global City.

In must be noted though that the GaWC rankings were always designed as a calculative methodology of urban connectivity, a measure of economic globalisation which, as Taylor (2004) theorised, is city-led, and have never purported to be a *dirigiste* of the urban condition *in toto*. However, Pandora's Box had been opened and the proliferation of city rankings on a whole range of different indices (including the rather arbitrary 'quality of life' variables) has continued. These more attributional rankings (in that they measure what is *within* a particular city

Figure 2.1 Toronto Global City.
Source: Author's photo, June 2012.

rather than the networks *between* cities which is the precondition of the GaWC rankings) have proliferated in recent years. There have been a whole range of institutions that have attempted to provide their own rankings of cities, each based on slightly different variables. Major financial institutions such as Foreign Policy, Mercer, PricewaterhouseCoopers and The Economist have all produced regular annual league tables of cities based on attributions ranging from pollution levels, property values and infant mortality rates. Being at the top brought with it a sense of gravitas that your particular city was a Global City, and hence part of the elite group of cities that, according to the ideology, 'commanded' the world economy.

Given the recent financial crisis, is it perhaps questionable to argue that the global economy is or has ever been 'controlled and commanded' (Smith 2014), yet the concept of the Global City as something to be achieved persists. However, as the term is used more and more by institutions and individuals who are unaware of its associated epistemology, it has lost its empirical and ontological 'baggage'. Indeed, as Smith (2013: 2296) has noted:

> What has been mobilized by practitioners (policy makers and planners) is an empty phrase – the 'world city' or 'global city' – devoid of its original neo-Marxist meaning and critical import against the very idea of the neoliberal

'world city' or 'global city' as a desirable goal, hollowed out to be no more than a relatively autonomous politico-economic agent and echo chamber for the interests of neoliberal globalization.

Hence, in a Baudrillardian simulacra, the representation of a particular city as a Global City has become the meaning in itself. Somewhere between a second- and third-order simulacra (Baudrillard 1994), the Global City is branded and hawked as an entity in itself, without any relation to the original methodological and theoretical connotations, but overall overtly aligns with the economistic mechanisms that brought it into being. Take, for example, Figure 2.1 above, taken from a hotel magazine in Toronto. The text reads 'A city known for its diversity, Toronto quite literally brings the world to its doorstep on a daily basis.' Notwithstanding the rather liberal use of the world 'literally', the term Global City is used in this context to denote the cultural diversity and hybridity that exists within the city of Toronto. In a throwback to Hall's (1966) initial description of a World City, the virtues of rich cultural diversity are posited as creating a Global City. Also, to go back to the rhetoric from Sydney, its official website details how the term 'Global City' is being used to denote the cultural diversity of the city. These colloquial examples are representative of the way in which the Global City term is infiltrating urban policy and promotional discourse in ways increasingly disconnected from its academic roots; it is, as Smith (2013) noted, being mobilised internationally, devoid of any meaning. It is worth noting, however, that the page on Sydney's website does state that 'the Global and World Cities group data inventory compiled by Britain's Loughborough University found Sydney is the seventh most connected city to the global economy' (City of Sydney 2013b: n.p.).

The concept of the Global City is of course not without its academic critiques. There is no need to cover these in intricate detail here, but it is worth outlining that they fall into two broad poles. One the one hand, the Global City is characterised by acute social polarisation (Sassen 2001). The APS firms that are housed in the Global City are populated by a 'transnational capitalist class' (Sklair 2001), who require an army of service workers to conduct the menial tasks that the elites do not. This creates what Castells and Mollenkopf (1991) noted as a 'dual city', with an increasing number of people at both 'ends' of the income spectrum. The other 'pole' of critiques is focused on how the Global City is highly imperialistic. It encourages a developmentalist discourse and creates a swath of 'Ordinary Cities' (Robinson 2002) that are characterised as somehow 'underneath' these elite Global Cities (something which is of course exacerbated by the constant (re) production of city league tables.

Despite these valid critiques, however, the Global City has now become part of the common parlance of urban policy and promotion. The increased pervasiveness of this term (and the other policy-friendly terms that will be examined throughout this book) is a symptom of the preponderance of neoliberal philosophies, and the shift from urban managerialism to urban entrepreneurialism (Harvey 1989). Cities are now managed like corporations with large marketing budgets, public relations departments and marketing strategies (as we saw with the Tel Aviv Global City unit). Such post-Fordist or postmodern urbanism has been characterised by a 'cultural turn' (Amin and Thrift 2002), in which urban strategies have used culture as a competitive tool. Deindustrialisation, the rise of BRIC countries and the availability of cheap labour in the sweat shops of Asia, Africa and South America, meant that for Western nations, the investment in cultural assets was not only desirable, it was essential for the maintenance of economic global dominance. Culture was becoming an important economic asset as it provided a means by which neo-imperialist economic relations (aka Walterstein's (2004) World Systems Theory) could be maintained. The term Global City then is so important in this regard as not only does it convey the image of a city being a critical node in the global economy, but it can also be a vehicle to denote the 'softer' cultural dimensions of a 'world city' (despite the original neo-Marxist roots of them steering well clear of such characterisations). These cultural dimensions, however, are typically a veneer or even pastiches of urban cultural life that are cherry picked to adhere to the cultural narrative being afforded to a particular city. Moreover, it is overtly the consumption of this culture that is championed, as in order to sustain a city's economic dominance, it needs to cater to all the consumption patterns of the financial industry elites whose practices maintain that influence, as well as house the lower-income workers who will service them (with such characteristics passed off as multiculturalism). In other words, the interjection of culture into a Global City paradigm actually justified the increasing commercialisation of a city's cultural assets as progressing the city's global competitiveness. As such, culture is infused into the Global City paradigm through an economic instrumentalism – a culture that can be packaged and sold:

> The potential for selling culture has been grasped eagerly by urban decision-makers after decades of uncertainty about how to meet the needs of an expanding urban population against a background of sustained urban deindustrialization and loss of economic rationale. (Amin and Thrift 2007: 151)

'Culture' now is being utilised linguistically by urban policies as a means to engender economic growth and urban development, and we see the raft of measures of urban regeneration predicated under the rubric of culture-led development (Evans 2005), more of which will be discussed in Chapter 4.

However, this particular rendition of 'culture' is ideologically bereft of any agency beyond economic stimulation. Moreover, it stands to reason that if culture is to be the new currency of economic globalisation, and if cities are to compete globally, then they need to be *seen* to be places of investment. It the age of urban entrepreneurialism, it is the image that is all-important. The postmodern-infused cultural and entrepreneurial turn has led to cities that want to compete, searching 'for quotation and fiction rather than invention and function' (Harvey 1989: 13). And what better way to compete internationally than to advertise yourself on the global stage? City branding is thus an important factor to consider here as it fuses the entrepreneurial actions of 'new' Global City activity with the facade of cultural provisioning. Hence, the ideas of city branding will be the focus of the next section.

City branding

In one of Sydney's busiest transport hubs, Town Hall Station, emblazoned across an interior wall is an eight-foot-high advert for Hong Kong. As you can see from Figure 2.2, it proclaims this former British colonial outpost as 'Asia's World City'. With a spectacularised image that is text-light, prominent, splashed with striking colours, memorable vernacular and in a prime visible location, it bears all the hallmarks of a direct advertising campaign not out of place in a top global multinational corporation. The 'Hong Kong Tourist Board' logo and text is just visible at the top of the advert and the website address gives as much information

Figure 2.2 Poster 'advertising' Hong Kong in Sydney.
Source: Author's photo, 15 April 2013.

as any passing person will need. Sydney and Hong Kong share key strategic links for many companies based in the Pacific Rim. Many Australian firms wanting to expand to China will settle in Hong Kong initially. It is also a busy tourist route in both directions. From a marketing perspective then, it is little wonder that the Hong Kong municipal government felt that spending valuable city resources on advertising space in Sydney would be worth it financially.

Given the rise of urban entrepreneurialism and the global competitiveness it has engendered, the race to become a World or Global City is exactly that – a race. To capture the global flows of people, ideas, capital and finance, cities need to stand out from the rest and are turning to marketing, PR and advertising executives to aid in this process. In the times before a deindustrialising international economy, it would have been the purview of trade delegations to visit other countries and cities to drum up foreign investment through interpersonal networks, inter-firm agreements and corporate mergers and acquisitions. However, in a globalised economy where culture is the critical factor in not only attracting international business but also tourists, cities need to be seen to be branding themselves in the advertising space of internationalised media. With the importance of the knowledge economy, the increased visibility of branding and its increasing importance to revenue, cities must market themselves globally. Urban entrepreneurialism and the rise of the Global City rankings has forced cities to spend more of their budgets on marketing, branding and PR. In branding cities in a particular way, though far from creating uniqueness, it increases the Lefebvrian industrial city through the homogenisation of cityscapes in advertising.

The most famous example of urban branding is undoubtedly the 'I love New York' slogan developed in the 1977 by Milton Glaser. Greenberg (2008) has detailed the way in which New York used branding, marketing and PR vehemently to redress the intense economic and social decline it was experiencing in the 1970s. The city was blighted by crime, unemployment and social degradation, and these characteristics were being catalysed by the images of the city in the ever-expanding global reach of films and TV shows. In short, the city was going through an image crisis. The urban governance structures throughout the 1970s and 1980s began to adopt neoliberal 'Reganomics' philosophies as they saw it as the only means by which the city could experience growth once again. But where New York excelled, according to Greenberg (2008), is that it coupled this with an aggressive, privately sourced marketing campaign to redress the city's international image as much as its socio-economic ills. The entire city's governance structure shifted to accommodate private companies that engaged in entrepreneurial and creative activity, and over time became a model of how a city can use branding and image manipulation to recover from a crisis. As such, the 'I love NY' logo which was appropriated from the New York state to New York City, has become one of the most iconic (and replicated) urban 'brands' of all time.

The processes involved in city branding are highly corporatised, and the deliberate manicuring of how a city is portrayed in the media becomes critical. It is also not just limited to cities in their entirety, the branding of particular boroughs, gated communities and specific neighbourhoods have also inculcated a marketing agenda to help 'sell' it to a wider, global audience. Greenberg (2008: 14) notes that 'one of the most important effects of branding is to naturalise, and reify, the image of its commodified object, and so to obscure the social and historical processes that produce it'. Austin (Grodach 2012), Turin (Vanolo 2008), Milwaukee (Zimmerman 2008), Montreal (Rantisi and Leslie 2006), Bilbao (Grodach 2010) and many others have all been shown, in one way or another, to create an image or brand in order to sell their city. But in so doing they create a pastiche that downplays the long economic, social and cultural histories. This is of course the remit of branding from a corporate perspective, to reduce to a singularity the complexities of a particular commodity (Aaker 2012). Neoliberal philosophies are imbued with the notion of the consumer as being highly influenced by marketing, PR and branding, and as such advertising plays a critical role not only as the propagator of neoliberalism, but also in the valorisation of commercialism as a means of growth. With increasing entrepreneurial activity in urban management, other cities and nations are viewed as markets to explore and consumers that can be 'captured'. It is therefore imperative that advertising and branding is part of that management process. Put crudely, if you are to run the city like a corporation, then you also have to sell it as one. Therefore the branding of cities has become an international economic activity in itself, with major advertising firms involved in creating urban marketing campaigns. The firm Oglivy & Mather, for example, are one of the world's prominent advertising companies, with a client list that includes Coca-Cola, Ford, BP, American Express and many other globally recognised 'brands'. To continue with the Sydney case study, in 2010, Oglivy & Mather created a campaign for the urban government called 'Sydnicity'. A television advert, posters, social media and glamorous events were all part of the campaign to market the city of Sydney around the world. More than simply a tourism campaign, the video commercial (which can be found online) is scripted as follows:

> *Try finding one perfect word to describe Sydney*
> *You won't find it in any book*
> *Maybe you'll find it in a melody or a song*
> *Or in the city and way it comes alive*
> *Perhaps you'll hear it whispered in the waves*
> *Or you'll find it somewhere set in stone*
> *A word for all the vibrant colours and flavours*
> *And the amazing views that take your breath away, everyday*
> *A word that really sums it all up*
> *Sydnicity. It's what makes Sydney so Sydney*

One of the interesting facets of this campaign is that the neologism 'Sydnicity' was riffed from words such as velocity, iconicity, synchronicity, electricity, authenticity, vivacity, tenacity and so on (words which flash on the screen toward the end of the commercial). The advert epitomises the late-capitalist, neoliberalised urban entrepreneurial strategies that cities are now employing, and bears more than a linguistic resemblance to the language used in the Creative City document discussed earlier in the chapter. Rather than focusing on a singular (industrial or cultural) facet of the city to be championed, it makes a direct play toward glorifying the tacit cultural and affective atmospheres of the city into a singular 'word'. The words they use – velocity, electricity, vivacity and so on – and the urban cultural offerings that are alluded to in the piece are demonstrative of the widespread shift toward the affective language of urban life within city marketing campaigns. It also explains the increasing usage of time-lapse video as a means of advertising the city given that it glorifies flow and mobility and hides friction, blockages and stickiness. Aesthetics that induce vitality, buzz and general excitement are in and of themselves devoid of any particular need for substantive meaning. But taking a more critical analytical stance, they are as about as useful as saying a city is 'full of life' (more is made of the 'language' of urban development in the next chapter). But strung together and juxtaposed with fast-paced, sweeping, picture-postcard shots of the city's commercialised cultural offerings, the campaign formulates a 'brand' that engenders a clean, mobile, efficient, commercial and high-culture-rich city. But a closer look reveals that the video depicts activities that are both stereotypical and limited to those with the social and financial capital to experience such activities. Classical music concerts, expensive restaurants with spectacular views, beach living and sailing – these are all activities that are afforded to the elite and/or tourists. Also, the deeply engrained racial tensions within the Australian community are not evident; the actors used in the video are overwhelmingly white and pictured as having large amounts of disposable income. The video also centres on the tourist 'hotspots' of the Sydney Opera House, the Queen Victoria Hall, the Botanical Gardens and the harbour; little is made of the city's sprawling suburban area to the West. The brand 'Sydnicity' is therefore an exercise in reducing the multiple, non-linearity and complex topologies of the city, in this case Sydney. It is hence very much Sydney's attempt to look for 'quotation and fiction, rather than invention and function' (Harvey 1989: 13). From a neoliberal and marketing perspective, it is an ingenious creative act as, while offering a bite-sized, easily recognisable and consumable slogan, it is pregnant with other more affective notions of the cultural offerings that appeal to foreign investment and tourists. But from a critical perspective, it espouses an urban entrepreneurialism that whitewashes the city of any tensions by neutralising it visually. Racial tensions, economic inequalities, social ills, environmental pollution, religious hatred – all the problems of urban life are ideologically suppressed.

Branding therefore is not only an important tool in neoliberalising the city and engendering a more corporatised aesthetic, it also has direct political agency as it subtly removes the images of a city's social problems from the public imaginary. It can be attributed as a Gramscian (1995) form of hegemony that is enforced through a 'common sense' logic, a soft cultural power that is not violently or oppressively implemented by an elite, but permeated through media, cultural institutions and other forms of pervasive knowledge (Jessop 1997). Such a view of hegemony is an important characteristic of the Creative City, indeed it is a critical one. It neutralises social issues that can be antagonistic and subvert the accumulation of capital. A Gramscian form of hegemony uses artistic, symbolic, semiotic and 'creative' means (such as city branding) to continually adopt the new, and as Mouffe (2013: 90) has argued, 'today's capitalism relies increasingly on semiotic techniques in order to create the modes of subjectivation that are necessary for its reproduction', something which Berardi (2012) has called 'semiocapitalism'. The desire of the Creative City to constantly renew its subjectivity is a critical (neoliberal) mechanism that it employs to ensure its progression and development, and hence another reason why the usage of creativity was the logical (common sense) progression of neoliberal urban development agendas. Branding a city as creative puts into ideological practice the hegemony of capitalistic gains. Its ingenuity comes from the presupposition of any resistance, subversion or anti-hegemonic actions as simply a 'creative' extension of the Creative City. Hence why it was a systemic requirement for the continuation of capitalistic gain as it created systems that more efficiently and readily appropriated forms of critique, resistance and subversion.

Branding therefore becomes a critical component of urban entrepreneurialism because of its ability to engender creativity as part of its remit. It speaks to the desire of cities to compete internationally, but also shows how they are rapidly commercialising their cultural assets. And it is this focus on (or 'turn to') culture that further foregrounded and indeed *necessitated* the emergence of the creativity paradigm and the Creative City itself. Urban officials took up the creativity paradigm so readily as it spoke directly to, and justified the practices of, this new desire of urban governments to commercialise its cultural assets. To remain competitive and to maintain a position atop the global hierarchies of cities (and preserve global hegemony), policies of creativity were critical as they justified the financial support to the development of cultural amenities, as well as the production of new ones. Many of the commentaries on this cultural turn in urban entrepreneurialism and the popularity of the creativity paradigm have highlighted this development (Peck 2005); however, there is a highly related process that is fundamentally critical to the eventual 'success' of the Creative City paradigm, namely that the concept of the 'creative industries'. Such a conceptual link is often implicit in the literature, but to date has rarely been critically explored. There is then a need to excavate the process of the inception of the creative industries, as along with neoliberalism, branding and the Global City, it forms a critical component to the narrative of the systemic need and eventual realisation of the current Creative City paradigm.

The creative industries

The 'creative industries' as a concept has been intensely debated. It is now global in reach with countries from Europe, Africa, Latin America and Asia implementing some sort of creative industries national policy. However, its inception as a concept can be traced back to the political landscape of Great Britain in the latter parts of the twentieth century. The 'Mapping Documents' publications of 1998 and 2001 from the UK government's Department for Culture, Media and Sport of the Labour government of the time have gone down in academic folklore as the bedrock of political vacuity on which the entire concept now rests. Since then, there have been books, reports, papers, academic departments and job descriptions, think tanks and government officials all with 'creative industries' in the title or strapline. How did they come into being though? And why are they so critical to the development of the Creative City paradigm? There have been a number of texts that have articulated the creation of the creative industries concept, notably O'Connor (2007) and Flew (2012), who detail the political positioning and shifts in terminology that pre-dated the creative industries. O'Connor (2007) details how it was the Greater London Council (GLC) in the 1970s and 1980s that first began to integrate culture and the economy into a formalised political discourse (i.e. a cultural policy). Previous to this, the cleavage between art and the market (in the traditional sense) was ideologically total (see also Hesmondhalgh (2005)). The GLC saw the free market as an appropriate means of distributing cultural goods and, more crucially, a move toward encouraging economic measures and commercial processes to artistic and cultural production. What is also worthy of note is the fact that it was the GLC, the governance structure of one of the world's Global Cities, that instituted this and not a national government, because, as Harvey (1989) has noted, it was (and still is) cities and not national governments that frontload political 'innovations' of this kind, namely the economisation and introduction of market capitalism into the cultural realm (see also Taylor 2013).

So with the creep of economic and financial processes into the mindset and frameworks of cultural producers, the term 'cultural industries' began to be used to represent a collection of industrialised practices that could be coagulated into a 'sector', thereby labelling activities such as film, music, art, theatre and other creative practices with an industrial and therefore politically malleable parlance. Again, it was the GLC, under the influence of the political economist Nicholas Garnham, that foreran a cultural industries mandate, arguing that 'most people's cultural needs and aspirations are being, for better or worse, supplied by the market as goods and services' (Garnham 1987: 25), quoted in Flew 2012: 23). The marketisation of culture in this way by the GLC was part of the wider entrepreneurialism of management structures that was more evident within the Global City paradigm as noted above. The capitalistic tendencies of management structures in this way saw cultural production become less guarded and marshalled by national policies of cultural provision, heritage and social equality, and more susceptible to the shorter, more locally sensitive but mass-consumption orientated cycles of market forces. The prevailing Conservative government in the UK at the time, with its aggressive privatisation policies strangled the arts and cultural funding in the UK, which added to the overall need for funds from the market. Without straying into the debates around cultural value (for an intricate and explorative discussion, see O'Brien (2013)), it is enough to say that the cultural industries soon became an arena in which a tidy profit could be made quickly.

The marketisation of cultural activity in the UK though was in many ways catching up and emulating the US, which many saw as the model of how to economise cultural production. Hollywood in particular and its ruthless intellectual property protection laws (MacCalman 2004) was the 'model' of how to profit from creative practices, and the massive tech giants of Silicon Valley provided evidence that digital and technological creativity could prove very lucrative. This is why, in the infancy of the New Labour government that came to power in the UK in 1997, Tony Blair, the then Prime Minister, set up the 'Creative Industries Taskforce' and their first 'task' was to assess the Hollywood model and see how it could be retrofitted to the UK's cultural industry landscape. Notice, however, it was the *Creative* Industries Taskforce and not *Cultural*. This change in parlance was critical. Notwithstanding the prevailing shift in capitalist systems to the 'information economy' and a post-industrial discourse of knowledge and innovation, the term 'culture', in the mid-1990s, still had political baggage that prevented the Treasury from assigning finances and other central government resources. It conjured up images of an 'old' Britain, one that the incoming New Labour government was keen to avoid. This is also partly because New Labour's success in the 1997 election was due in large part to the *Cool Britannia* campaign, which saw Blair rub shoulders with some of the UK's most prominent artists and musicians. Blair was often photographed with Noel Gallagher, Tracy Emin and other prominent British cultural producers. It was a forward-thinking, 'cool' and creative governmental narrative, one that could capitalise economically from the increasing marketisation of the cultural industries. As Flew (2012: 14) notes:

> [The] creative industries as a concept was consistent with a number of touchstones of the redefining of the British Labour Party as 'New Labour', as it was spearheaded by Tony Blair and his supporters within Labour, with its recurring concerns with economic modernisation and Britain's post-industrial future. Its focus on the role of markets as stimuli to arts and culture was consistent with the notion of a 'Third Way' between Thatcher-era free market economics and traditional social democracy, that was nonetheless more accommodating of the role of markets and global capitalism than traditional British Labour Party philosophy and doctrine.

So, it follows that one of the first instructions by Blair was to set up a Creative Industries Taskforce, as it forced a change in the political vernacular from *culture* to *creativity*, but with little to no change to the actual industrial production that it labelled. The prominent industrial policy commentators (and think-tank directors) at the time, including Will Hutton, Charles Leadbetter, John Howkins and the Creative City eulogist himself, Charles Landry, were using the term as a means of describing the skillset needed to prosper in the new information, knowledge, digital and/or 'weightless' economy. The influence of think tanks in constructing governmental policy (forged from existing social and interpersonal networks) is another facet of neoliberal entrepreneurialism that increasingly characterises national and urban governments; so it follows that the language used by these institutions infiltrates into governmental policy (Peck and Tickell 2007). The parlance then of creativity also had the added and very important benefit of making the Treasury more amenable to releasing funds (Oakley 2004). The Taskforce, led by the then Secretary of State for Culture, Media and Sport, Chris Smith (now in the House of Lords), visited prominent industry leaders in the US and Hollywood, and began to 'map' the creative industries in the UK. In a very short timeframe, the DCMS published the taskforce's findings in the first Mapping Document in 1998 and its subsequent follow-up in 2001. In it, they outlined 13 subsectors within the creative industries, including among others film, software, arts and antiques, music, publishing and advertising. It is clear from the documents that it was a rather random, 'finger in the wind' exercise, with many of the subsectors containing very little 'creative' activity or production. Arts and antiques, for example, was measured exclusively via market sales, and indeed the very notion that an antique can be manufactured contemporarily is rather counter-intuitive to it being a 'creative industrial' activity. The inclusion of many tangential activities and industries such as software and advertising, the 'creative' elements of which were rather difficult to isolate quantifiably, bloated the figures somewhat and created an artificially productive (yet political and economically convincing) set of economic measures.

What was perhaps more critical to the development of the concept as a political economic tool was the definition that the Mapping Documents put forward. It suggested that the creative industries were 'those industries which have their origin in individual creativity, skill and talent and which have a potential for wealth and job creation through the generation and exploitation of intellectual property' (DCMS 2001: 4). It is perhaps the second part of that definition that had the most traction politically. The exploitation of intellectual property was the principle lesson from the US, and as such was a key driver in the formulation of the collectivised notion of the creative industries. The definition itself has been the subject of intense scrutiny and debate (see Galloway and Dunlop 2007), critiqued as valorising the market-led approach to cultural production and defenestrated from the political mantra any sense of cultural meaning that does not translate into economic benefit. In a sense, what the political economic formulation of the creative industries did was to make cultural production politically acceptable through ideologically stripping it of any public, civic or social value. If *all* the returns to investment in cultural production yielded a financial reward, then such investment could be easily justified within national government treasuries. The results of such a reconceptualisation of cultural production can be seen in Figure 2.3 which shows a UK Trade and Investment (UKTI) promotional poster from the 'Great Britain' campaign. The campaign has posters all over the world, with a number of different slogans designed to promote the UK's strengths such as 'Business is Great', 'Shopping is Great' and 'Creativity is Great', and in the case of Figure 2.3 'Culture is Great'. The fact that it is a promotional arm of UKTI aimed at exploiting the UK's cultural assets to attract investment exemplifies how culture is being rearticulated as an economic stimulant.

Figure 2.3 UK Trade and Investment promotional poster in New York, USA.
Source: Author's photo, 25 February 2012.

In the post-Fordist 'new' industrial society, knowledge, digitisation, intellectual property and information are the monetary drivers, and so in order to maintain a politics of growth and economic development (which as Harvey has detailed for the last quarter of a century is inextricability linked to the urbanisation process), there is a political and economic necessity to monetise as much 'new' economic production as is possible – to engage in neoliberal policies that are adaptive, agile and have assemblage qualities. Put bluntly, in order to maintain itself, capitalism needs to build cities, and to do so it needs to maximise profit from the production process, which is why the shift in vernacular from cultural (which denotes a public and civic value) to creative (which is far more entrepreneurial in its articulation) was not only politically savvy, it was another *systemic* requirement of the political economic process of capitalist urbanisation.

Summary

The *language* of the Global City (stemming from its academic formulation), couched in neoliberal philosophies and entrepreneurially friendly governance structures, has catalysed privatised urbanism and inter-city competition, and the inception of the creative industries has redirected the production of culture into market-led capitalism. This is why the creative industries as a politically constructed concept was crucial to the predilection for a post-deindustrialising global economy and hence to neoliberal urban governance more broadly. It offered a politically palatable and business-friendly framework of cultural consumption. Moreover, these articulatory 'innovations' are related through the general postmodern turn of neoclassical economics, yet their 'sharp ends' dovetail together in tandem to set the foundations for the Creative City. If a city is to compete internationally then just like a corporation it needs to be innovative and exploit a market (via branding). And now, as cultural and creative production (aka the creative industries) is politically needed to fuel capitalistic urbanisation, then a specific urban governance system *had to* perpetuate. Hence the idea of the Creative City was the next logical step within a neoliberalising ideology. The ingredients of rebranding and marketing cities (wrought with the market-leaning ideologies of the Global City) and the political desire to adopt the creative industries as an industrial policy were in place. And as neoliberalism requires local specificities to reinvent its philosophies (Springer 2010), then the geographical and locational ingredient was all that was needed to realise the neoliberalisation of creativity to its capacity. Therefore, it becomes clear how the Creative City was born from a perfect storm of urban and economic political processes and changes in urban managerial rhetoric, and hence was a *systemic requirement* of the latest iteration of neoliberal urbanisation.

However, there is another process, another creativity narrative, that is so far missing from this excavation of the Creative City paradigm – that of the *creative class*. If increasing entrepreneurial urban governance machinery provided the political will in which creative industrial practices can proliferate urbanisation, then the creative class theory gave cities the political language they needed to mobilise their neoliberal policies free from any competing political resistance. Given the importance of the creative class theory to the Creative City paradigm, it requires a great deal more effort and focus to unpack it. Hence, the next chapter will show exactly how the uptake of the creative class theory mobilised the Creative City paradigm to become ever more dogmatic in becoming the urbanising force that it is today.

3 The Creative Class(ification) of Cities

The Stokes Croft area of Bristol is the city's cultural and creative hub. The People's Republic of Stokes Croft (PRSC) are a grouping of local Bristolians who have championed the alternative, anti-hegemonic milieu of the area. It is primarily a residential area, but the main road of Stokes Croft itself has many shops, cafes, pubs, boutique galleries and artist studios, some of which could be considered to be an 'alternative' form of cultural provision, something that the PRSC are keen to promote and maintain. Below is an exert from their mission statement:

> PRSC will seek to promote creativity and activity in the local environment, thereby generating prosperity, both financial and spiritual.
>
> PRSC will work in all ways to enhance the reputation of Stokes Croft as a globally renowned Centre for Excellence in the Arts, both in its own actions and by encouraging the action of others.
>
> PRSC believes that the strength of the local Community resides in its creativity, tolerance and respect for each other.
>
> (People's Republic of Stokes Croft 2011: n.p.)

As can be seen from this and from other parts of the statement, the PRSC are passionate about maintaining a certain 'vibe' in the area, one of creativity that has 'spiritual' as well as financial elements. Tolerance is also noted as being an important characteristic of a strong community, and a quick navigation through their site highlights a distinct politics of activism and anti-establishment practices. By working with but mostly against city council forces, the PRSC have battled to maintain these principles, but with varying degrees of 'success' (depending of course, on how you qualify that success). For example, the area is (in)famous for, among other things, the large murals and street art on the external and internal walls of the buildings (for example, see Figure 3.1). These are often political in nature (see Figure 3.2), but the practice of street art and adorning the facades of local shops and houses with large, brightly coloured murals is tolerated in Stokes Croft far more than it would be in other more 'mainstream' parts of Bristol. These have gained international notoriety and the area is now a popular tourist destination for street art enthusiasts.

Figure 3.1 A piece of street art in Stokes Croft, Bristol.
Source: Author's photo, 22 September 2011.

Figure 3.2 Political street art in Stokes Croft.
Source: Author's photo 22 September 2011.

The founding of the PRSC has been predicated upon the area's history of social unrest and anti-authoritarianism, dating back to the 1980s when there were riots over what residents saw as institutionalised racism in the local police (Clement 2012). However, more recently in 2011, there were nights of rioting triggered by the persistence of a large supermarket chain wanting to open a store in the area. Tesco, as a brand, has become a byword for banality and homogenous urbanisation in the UK. The presence of a Tesco is often characterised (with some justification) as the loss of a place's identity, a barometer of the level of independence or 'authenticity' that a locality has and an indicator of gentrification. Put bluntly, if your area has a Tesco, then it has sold out. As such, given the counter-cultural milieu that has been carefully manicured by the local residents (and the People's Republic) over the years in Stokes Croft, the presence of a Tesco mini-market was viewed as going against the grain of the area's ethos and was fiercely campaigned against (see Figure 3.3).

Also, Stokes Croft is known for its squats and homeless population. The St James Barton roundabout is often referred to as the Bear Pit and is a common location for homeless people in the city (Kiddey and Schofield 2011). In April 2011, 160 or so riot police raided 'Telepathic Heights' (one of the more 'permanent' squats in the area) on a tip-off that residents were planning to petrol bomb the recently opened Tesco that is located across the road. The sheer scale of the police presence was enough to enrage residents of Stokes Croft and the conflict scaled up that night with looting of the Tesco and other corporate institutions and rioting in the wider area of St Pauls (Clement 2012).

Since the riots, however, the area of Stokes Croft has seen no reduction in the process of capitalistic development, with luxury flats erected on the sites of former squats, landlords starting to increase their rents and the slow colonisation

Figure 3.3 Anti-Tesco graffiti in Stokes Croft, Bristol.
Source: Author's photo, 22 September 2011.

Figure 3.4 The gentrification process in Stokes Croft, Bristol.
Source: Author's photo, 22 September 2011.

of informal spaces by formal and commercial institutions. Figure 3.4 neatly exemplifies just such a process: a newly opened commercial coffee shop chain in a landscaped and beautified area, with a sign advertising newly built student and executive flats obscuring one of the squats in the background.

In a paradoxical process, as a direct consequence of the area's alternative ethos, the anti-commercial and subversive activities that thread through the day-to-day rhythms of the locality, Stokes Croft is (unfortunately or not, depending on your point of view and your postcode) gaining the characteristics the very ethos is guarding against. More affluent, bohemian and 'creative' people and professionals are moving in as they see the area as stimulating an alternative cultural milieu, one that is befitting of their lifestyle. This though drives up rents, and landlords begin to see the investment potentials and buy out squats and small art studios that cannot afford to pay the now higher levels of rent. These are then often demolished with executive houses and flats built to accommodate the growing interest in the area – the area is being gentrified. Before going into why this is catalysing the influx of the creative class and contributing to the Creative City paradigm, there is a need in the first instance to contextualise gentrification as part of the ever-changing 'language of urban growth' in city development over the last few decades.

The language of urban growth

On 29 September 2012, Professor Neil Smith of the Graduate Centre of City University of New York died at the age of 58. He had a profound effect on urban geography and his pioneering critical work on the entrepreneurial style of city development (and his tragic untimely death) only served to ossify the vigour of his many colleagues and like-minded contemporaries in their denouncement of this entrepreneurial form of urban growth. One of Smith's major interventions into the urban geography literature was the unpacking of gentrification and the way in which it is (the urban) part of a wider structural change enforced by the increased uptake of capitalist economic policies. The word 'gentrification' was coined by Ruth Glass in 1964 when she documented the struggles of London's working class against increased house prices, the development of old housing stock, the shifts from rental to ownership and the increasing displacement of the urban poor from where they had built communities (i.e. the city spaces being commandeered by the elitist classes as per the historical situation in rural environments with the landed gentry – *gentri*fication). Nearly three decades ago, Smith (1987: 463) built upon this work when he noted that:

> The crucial point about gentrification is that it involves not only a social change but also, at the neighbourhood scale, a physical change in the housing stock and an economic change in the land and housing markets. It is this combination of social, physical, and economic change that distinguishes gentrification as an identifiable process or set of processes.

Over the intervening years, gentrification has become systemic to the way in which cities have developed as a part of the neoliberal assemblage. The increased intensity of urban entrepreneurialism has the gentrification process enmeshed within its very fabric. As Slater notes (2011: 572) 'gentrification commonly occurs in urban areas where prior disinvestment in the urban infrastructure creates opportunities for profitable redevelopment'. This is something Smith outlined with the Rent Gap theory (Smith 1987), which is (simply put) the financial difference between what an area of land is worth pitted against what it *could* be worth were it to be built upon. Private interests looking to maximise profit from development then look for the largest possible gap between the acquisition of the land and the final sale price once they have built upon it. As such, urban governance strategies are always engaging with sites that are in (perceived) decline, states of financial and social depression or with a dilapidated housing stock, looking to develop them into economically functional (and hence profitable) parts of the city. If this process occurs in places where people live and then are subsequently displaced, then this is the social detriment of gentrification, but is often viewed by the instigators and investors as merely an unfortunate by-product. As Slater (2011: 572) again argues, gentrification occurs 'where the needs and concerns of business and policy elites are met at the expense of urban residents affected by work instability, unemployment, and stigmatisation'. This diminution of existing communities and societies is a clear disadvantage of such developments, one that the instigators do not seek to avoid (as that would mean a drop in profit), but seek to make 'invisible' through the contortion of the rhetoric and the everyday lexicon of urban growth.

To exemplify this with perhaps one of the most brazen and outlandish developments of recent years in the UK, one only has to look at the recent Liverpool

Waters development. In March 2013, the project was given the green light by the UK government without going to public inquiry (Smith 2013). Eric Pickles, the then Communities and Local Government Minister, 'waved through' the project citing the intense need for income into the area, which has stagnated due to the current prevailing financial difficulties. And perhaps more tellingly, Peel Holdings (the investors of the project, who also financed, manage and own MediaCityUK in Salford, of which more detail will be given in the next chapter) threatened to pull out if a public inquiry was to go ahead, citing that it would incur unnecessary expenses and a delay to the project (Harrison 2014). The current austerity programme of the UK government has seen drastic reductions in finance available for housing and social services, and so when private investors offer (in this case) £5.5bn of investment it is unsurprising to see the UK and Liverpool city government welcome such an investment and even bypass public inquiries to do so, such is the influence of Peel Holdings. Albeit on a larger scale, this kind of development project is typical across cities around the world, with private investment prioritised over social and cultural issues, made all the more 'invisible' by the language of the rhetoric that is put forward by the developers. For example, the following is taken from the website for Liverpool Waters:

> The Liverpool Waters vision involves regenerating a 60 hectare historic dockland site to create a world-class, high-quality, mixed use waterfront quarter in central Liverpool. The scheme … will create a unique sense of place, taking advantage of the sites' cultural heritage and integrating it with exciting and sustainable new development. (Peel, n.d.: n.p.)

As is typical with the rhetoric of such developments, the language is one of revitalisation and regeneration, negating the gentrifying properties of the project. Liverpool Waters will be built mainly upon 'derelict' dock land, but closer scrutiny of the plans and the current usage show that there are businesses located in the docks that will be compulsorily purchased, as well as inner-city housing stock that will be hugely effected by the development. These may be seen as inconsequential to the larger development, but the fact that the aesthetics of the Liverpool Waters project has seriously jeopardised the UNESCO heritage status of Liverpool's waterfront is a major point of contention for the project's opposition. Counter-arguments from the local business community (and of course from Peel themselves) note that the heritage laws that that are designed to preserve the architectural aesthetics of Liverpool's waterfront are archaic and block progress. The dualistic argument of 'adapt or die' is put forward as a way of justifying such a massive change to the city (Brown 2012). However, such an argument belies the broader issues of the Right to the City (Harvey 2003; Marcuse 2009) as they forgo current and future civic, urban and social engagement for short-term financial reward and developmental lexicon. The genuflection of state officialdom to the financial might of private investors creates an urban environment that is solely representative of a political and financial elite, with little room for compromise. For Liverpool Waters, this is shown by the refusal of Peel and the wider supportive urban governance system to change the plans in light of the damning critique by UNESCO, instead holding fast to the desire to mould the urban form in the image of their own choosing (Harrison 2014). The argument about Liverpool Waters will continue long into this century (not least as Peel have indicated that it will take 50 years to complete) and it is foolhardy to assume any peremptory assessment at this early stage. However, in terms of exemplifying the gentrification process, it highlights how the urban is the physical manifestation of a power struggle, in which the financial powers play with the language of 'development' in order to circumnavigate and 'hide' the inevitable gentrifying properties of their schemes.

So far in this book, for shorthand ease (and for want of a better phrase) I have been using the word 'development' to characterise the growth and general change to cities. The *re*development of cities then logically refers to how particular spaces that once had a particular use but have now declined are being changed once more. This parlance though has a subtle yet profound mechanistic quality of rendering the process politically and geographically neutral. Other terms have also begun to enter the policy realm, such as 'urban renaissance', 'regeneration' and 'revitalisation' (Lees 2003b). As has been argued, how could you not want to be part of an 'urban renaissance'? Also, the subtle use of the re- (*revitalise*, *re*generate, etc.) implies a state of dereliction that requires addressing. For example, to '*revital*ise' implies that the area was not vital in the first instance. To 'regenerate' infers that the area is in a decomposed state that needs to be nurtured back to a state of acceptance once more. In some cases, these are sites of industrial dereliction, where the landscape is popularly characterised by general ruin, inactivity and economic stagnation. In others, the area has a low-income housing stock, characterised as blighted, anti-social and a drain on resources. In either scenario, the message and vernacular from the developers and investors is one of 'upscaling', of creating a better and more vibrant place to live and/or work, and hence is prime for investment and subsequent *re*development. The vehement critiques of the gentrification process point to the way in which the needs of the local communities (residential, business, environmental and cultural) are often removed from the political process. Public consultation laws being overridden or public inquiries not being granted are justified because they would hold up or stop altogether the (much-needed) financial rewards that comes from the private investors. Such debates are played out throughout the urban geographical literature, and while inherently critical to the contextualisation of the infrastructural policies of the Creative City, cannot be rehearsed here. But as has been noted so far, the whole Creative City ideology stems from the need for an alternative policy 'fix' (Peck 2005), one that encapsulates the 'new' knowledge-based economic parlance, one that can continue to justify urban development within such a linguistic framework and one that also plays into the political and economic putative (and increasingly dogmatic) 'creativity' paradigm. In touching upon the gentrification arguments (and the change in language that is enacted to attempt to circumnavigate the social and cultural issues), it is clear that the Creative City very much incorporates this gentrification process as part of its political economy as it is seen as creating desirable places for the creative class to live, work and play. But who, or perhaps more accurately what, is the creative class?

Creative Class

Stokes Croft has experienced an influx of what have been deemed the 'creative class' (something which Liverpool Waters will no doubt experience too). To understand what this term means and how it has become so critical to, and linguistically allied with, the notion of the Creative City, there is really a need to visit the work of only one man, Richard Florida. His seminal text *The Rise of the Creative Class* has been the subject of a multitude of debates, discussions, critiques and counter-critiques since its publication in 2002 (which was comprehensively revised a decade after in 2012 with *The Rise of the Creative Class: Revisited*). It seems that no urban geography publication has caused more controversy in recent times given the plethora of newspaper articles, podcasts, blogs, webzines, journal articles, academic conferences, monographs, edited collections and even parody Twitter accounts set up in light of the themes of the book. In it, Florida sets out his thesis of the 'creative class', who he defines thusly:

> The distinguishing features of the Creative Class is that its members engage in work whose function is to create 'meaningful new forms.' I define the Creative Class by the occupations that people have, and I divide it into two components. What I call the *Super-Creative Core* of the Creative Class includes scientists and engineers, university professors, poets and novelists, artists, entertainers, actors, designers and architects, as well as the thought leadership of modern society: nonfiction writers, editors, cultural figures, think-tank researchers, analysts and other opinion makers … Beyond this core group, the Creative Class also includes 'creative professionals' who work in a wide range of knowledge-intensive industries, such as high-tech, financial services, the legal and health care professionals, and business management. (Florida 2012a: 38–9, original emphasis)

These people, Florida argues, are fuelling the 'new' economy. The political conceptualisation of the creative industries outlined in the previous chapter, catalysed by think-tank-infused governance structures, put 'creativity' firmly on political horizons, and Florida's thesis delivered a way in which to epistemologically rationalise creativity as a human but more importantly financially viable trait. The creative class are the people whose economic actions are contributing most to growth as 'creativity has come to be the most highly prized commodity in our economy' (ibid.: 6). These specific groups of economic actors are therefore critical to the development and economic growth of a country, a region or a city. Moreover, the (seemingly American-centric) creative class are highly mobile and 'tend to move around different parts of the country; they many not be "natives" of the place they live, even if they were American-born' (ibid.: 58).

They are therefore more footloose than workers in other less creative professions, and as such are attracted to particular places and cities that offer a certain lifestyle that is befitting of their creativity, rather than choosing where to live based on the location of their job. As a result, one of the central tenants of neoclassical economic growth theory Florida turned on its head – instead of people following jobs, employers and the job opportunities they represent are now beginning to follow the people. Also, more than identifying the drivers of contemporary capitalism and challenging the traditional notions of regional and urban development, Florida argues that for particular places to succeed in the new economic landscape of creativity, to be attractive to the creative class, they need what he labels 'the three Ts' – technology, tolerance and talent. Technology refers to the level of high-tech industry prevalent in a particular place or city and other instrumental devices such as broadband provisioning. Tolerance, Florida argued, could be quantitatively conceptualised through indexes of the number of bohemians and gays in a particular city. He argues, 'a significant body of independent research confirms that openness to gays and lesbians is associated with higher levels of regional as well as national economic performance' (Florida 2012a: 274), and as such tolerant cities attract the creative class and hence perform better economically. As well as gays, bohemians are also a good indicator of a tolerant community, a group which Florida defines as authors, designers, musicians and composers, actors and directors, craft-artists, painters, sculptors and artist printmakers, photographers, dancers and artists, performers and related workers (Florida 2005). And another variable that is correlated with the tolerance of an area is the ethnic mix, or what is called the 'melting pot index'. Finally, the talent is needed. It is measured as the percentage of residents with a higher educational and/or bachelor degree. As the 'carriers of creativity', talented people attract the high-end jobs, economic restructuring toward the new, knowledge-intensive industrial output and, with it, urban prosperity. Having the three Ts in equal measure is a sure-fire way to attract the creative class, and therefore economic regional and urban success.

Stokes Croft in Bristol, given the edgy, bohemian, alternative and creative milieu that the PRSC have engendered so precisely is very typical of the kind of environment that Florida sees as attractive to the creative class, hence why in recent years, the area has seen the subtle shifts in the urban aesthetics toward those that are indicative of the 'settling' of the creative class (see Figure 3.4 for example). It is experiencing creative class*ification*. Academic literature, political documentation, media commentary, journalistic articles and online 'op-ed' pieces are all awash with other examples of cities and urban neighbourhoods that have experienced similar creative class*ification*. Moreover, there are stratospherically large amounts of literature that attempt to measure the creative class of this city or that region. Florida's thesis gained gargantuan political traction in the subsequent years after 2002, with countless cities and national governments across the globe hurrying to book Florida for speaking engagements in which he would highlight just how cities can succeed in the new economy through harnessing creativity, focusing on the three Ts and attracting the creative class. Armed with this new paradigm of growth that was based on academic work, new entrepreneurial urban governments focused on creating a globally competitive image that championed their culture and creativity assets. We have already seen in Chapter 2 how cities have used creativity as a means to promote themselves internationally, and with the uptake of the creative class thesis as a way of justifying such a promotional strategy, the creativity lexicon had become the most important urban policy tool of the twenty-first century.

However, there are as many (if not more) journal articles, books, blog posts, op-eds, newspaper and online pieces that counter Florida's arguments as there are that valorise the creative class as an urban development tool. The critiques are multifaceted, but their excavation and exploration are important so as to highlight how the Creative City is predicated upon a notion that has numerous and substantial problems. Such narratives question the calculative nature of the thesis itself, including the evidence it is based upon, debating whether the creative class really does stimulate economic growth or if their in-migration is a consequence of it, or indeed if the creative class really 'exists' at all but is rather a collection of occupations without any collectivised politics. As was discussed in the previous chapter, the Global City paradigm embellished an urban entrepreneurialism that has characterised contemporary urban governance. Similarly and no doubt relatedly, the uptake of the creative class theory is equally mechanistic of a shift to urban entrepreneurial and corporatised strategies. Therefore the following section will detail the critiques step by step, highlighting the problematic application of the creative class within urban politics.

Critiques of the Creative Class

Given the popularity of the creative class as a concept in urban politics, it was always bound to come under scrutiny, as is typical with many other theoretical and/or academic ideas that garner popularity beyond their immediate disciplinary and academic realms. However, the stratospherically rapid uptake of the ideas within city governance networks (city councils, private contractors, think tanks and the like) meant that these ideas were being mobilised and realised at an incredible rate, with tangible outcomes affecting not only the aesthetic, cultural, economic and political landscape of cities all over the world, but there was also a profound social impact upon the people living in those cities. As Gibson and Kong (2005: 551) noted, 'the desires of academics to become figureheads for certain concepts and to be heard beyond the confines of academia' is a 'noble enough sentiment'. Indeed, academia has often been derided for not being applicable to the modern world, and so when academic ideas are transferred into mainstream policy, the efforts need to be applauded. However, the major problems with the creative class concept has been (and continues to be) that not only is it critiqued for being based on shaky empirical foundations, it is also, perhaps more importantly, the catalyst for the increased economisation of culture within the urban context and has contributed to the deleterious effects that urban entrepreneurialism has on the social and cultural fabric of cities. It would be impertinent to suggest that there is a set 'typology' of critiques to the creative class, but given the vast literature devoted to it, and the desire to account for it all thematically, there is perhaps some utility to be gained in categorising the main strands – *definitions, empirics* and *politics* – and then extrapolating from there how these critiques offer an insight into just why the creative class theory can be problematised within an urban theoretical framework.

The, creative, class. Three loaded words conceptually which, when unpacked, cut to the heart of the definitional problems. Starting with *class*, the term has a complicated and protracted etymology within the social sciences. Fredrich Engels penned *Condition of the Working Class* in 1844 about his short time in Manchester working at his father's cotton mill. Witnessing the dire living conditions and increased mortality rates of the factory workers, he argued that the industrialisation and urbanisation processes of the time that came to characterise the Industrial Revolution was the main causal factor. In a damning narrative of the industrial-city complex of the then sprawling Manchester, Engels surmised that capitalist production necessitated a dualistic urban structure that 'condemned workers to a social suffering on an epic scale' (Slater 2013: 371) while the factory owners profited from such conditions. This tale of the social problems associated with the rapid industrialisation of Manchester provided the backdrop to Engels' famous collaboration with Marx, and from that the class system became part of the everyday parlance within social science inquiry. 'Class' therefore, denotes a certain socio-economic characterisation, political (im) mobility and a way of life that draws together the social disempowerment of the urbanisation process, the economic development of a capitalist system and the rapid growth of urban inequalities. The term has immense etymological and academic baggage that is difficult to extricate without risking generalisation or, worse, committing some sort of theoretical chicanery. The use of the word 'class' to describe a collection of occupations then, on the face of it, is in keeping with the capitalist production narrative of class. Moreover, Engels and Marx's notion of class analysis implies a common mode of production, and again with the creative class there is a sense these people could share a common productive ethos. However, there is an arbitrariness to the creative class categorisation that belies any common socio-economic goal, or indeed any ontological history. Pratt (2008: 110) argues that 'in Florida's work, class is reduced to taxonomy; moreover, one whose boundaries are not clearly defined ... Florida's occupational list is eclectic to say the least.' The grouping therefore of a seemingly random set of occupations (much like the randomness of the creative industry subsectors) is an artificial construct of a perceived set of consumption patterns aligned with a bohemian lifestyle. Krätke (2010) continues this line of argument by suggesting that the creative class follows a 'superficial' definition of class structure, one that is based on a 'mainstream sociological concept of class' (ibid.: 836) and not on the more tortured political economic reading that includes the hegemonic powers and the systemic social inequalities inherent within. As Anderson (1974: 76) noted, 'the study of social class must be based on social inequality'.

It is worth noting, however, that such inequality is intimated within the creative class thesis. Florida (2002: 71) argues that the presence of the creative class in any one city also requires the presence of a 'service class':

> Members of the creative class, because they are well compensated and work long and unpredictable hours, require a growing pool of low-end service workers to take care of them and do their chores. This [service] class has been created out of necessity because of the way the Creative Economy operates.

Mirroring the critiques of the Global City (Castells and Mollenkopf 1991), Florida's thesis entails that the people who 'do the chores' of the creative class are 'stuck for life in menial jobs' (Florida 2012a: 47), earn significantly less than those in the creative class and account for an increasing percentage of the (American) workforce. However, what the creative class ideology fails to account for is the systemic notion of social inequality within the very economic activities that it engenders. A more nuanced analysis of the notion of class would suggest that the very existence of a class structure is predicated upon the deliberate diminution and depression of those who support it. The creative class then, more than being predicated upon the importance of knowledge and creativity within the new economy, 'exists' in so far as it demands and subsequently exploits a service class. Therefore, more than simply being a response to the 'growing demands of the creative economy' (ibid.: 47), the rise of the service class can be attributed to the broader socio-economic conditions characterised by the neoliberal Global City and Creative City governance. As such, the creative class thesis as it has been presented so far, fails to adequately account for the inherent mechanisms that produce a service class. Moreover, members of the creative class and the service class share work–life attributes unaccounted for in the thesis. Despite occupying polar opposites of the labour market, service workers (particularly those in 'low-end' jobs that seem to be more prevalent in retail, hospitality and tourism) and members of the creative class share the more problematic qualities of precariousness, temporariness and employment insecurities (Ross 2008). Interns, temps, extras, freelancers, sole traders – these all exist within Florida's creative class categorisation yet have very similar problems. During previous ethnographic research I worked as a temp for a Sydney-based television production company. I had the luxury of being financially supported by my institution, yet many of the other workers I came into contact with in that brief period were working for free as they saw it as a way 'into' the creative industries. Living subsistence lifestyles in cramped temporary accommodation, many of these workers believed that continuing to work in such precarious conditions was a necessary step to gaining the all-critical lucrative career in television. The literature on the precariousness of work suggests that such characteristics are apparent in all aspects of the economy, but they appear to be particularly acute in the creative industries (see Bain and McLean 2013; Banks and Hesmondhalgh 2009; Gill and Pratt 2008; Hracs 2009; McAuliffe 2012; Ross 2008). Such flexible work (including ad hoc freelancing, zero-hours contracts, volunteering, unpaid internships and so on) is rife in the Creative City, yet the creative class thesis does not contain any mechanisms to deter or account for such issues.

Florida does offer a more instrumental solution, however; 'every job can and must be *creatified*' (Florida 2012a: 388, original emphasis). In other words, the creativity inherent in service jobs needs to be unleashed by 'tapping into the innovative and creative capabilities of service workers and engaging with them more fully' and this 'will ultimately make them more productive' (ibid.). Paying more for the services they offer will build a 'stronger middle class, enhance social cohesion, and create the demand that can help drive the economy forward' (ibid.). In sum, for the service class to work their way up the socio-economic prosperity ladder, they need to be recognised for their creativity. But if we continue to adhere to a creative class thesis, one that valorises the creativity of only a select few occupational groups, how can this ever be done? This leads onto the second problematic word of the definition, *creative*. The creative industries (as was seen in Chapter 2), the creative class and the creative economy (the nebulous notion that is an ideological coagulation of the two) are all *defined* by a creativity that adheres to a narrow remit – namely an economic one. If we are to celebrate the creativity of the service class, or indeed the creativity that is inherent in everyone (as Florida suggests is the case), then this requires an antithetical 'letting go' of a political economic definition of creativity. Krätke (2010) details how creativity within the creative class definition is limited to a set group of occupations with no devolvement into their working practices. Indeed, as Markusen (2006: 1922) has noted, 'he [Florida] conflates creativity with high levels of education.' Also, within the creative class definition, 'creative professionals' includes an amorphous grouping of consultants, technicians, firm managers, financiers, realtors and as well members of a political elite including public administrators, politicians, 'thought' leaders and members of what has recently been labelled in popular discourse as the 'commiteriat' or, more disparagingly, policy wonks. This group Krätke (2010) articulates as the 'dealer' class (but whether they are indeed a 'class' is open to the same debates of course), and far from contributing to economic growth, they are a threat in that they instigate financial crises (as was exemplified in the recent recession). Financial innovations such as subprime mortgage lending and payday loan companies have been highly detrimental to urban economic prosperity and have done a great deal to maintain, if not deepen, urban poverty and a dependence on high-interest loans. Yet within financial industrial rhetoric, they are seen as product innovations, as creative. Also, by including a political class and politically motivated commentators, think tanks and lobbying groups, the creative class ideology neutralises and depoliticises many of these activities. By bundling them in inconspicuously with high-tech innovators, artists and scientists, the politics of their actions are negated. Yes there may be creativity in the way that a think tank produces a report, but the politics of that report may well be such that they reinforce existing urban inequality or lead to policies that suppress economic growth (such as austerity). The question is not whether these activities are creative or not (although this is a line of argument I shall return to later in the book), but that within the creative class thesis creative practices are not identified.

Also, Florida argues that creativity is a resource found in everyone yet continues to eulogise that it is only economically viable within a certain group of people, i.e. the creative class. Wilson and Keil (2008: 841) have counter-argued that 'the real creative class in these cities is the poor', specifically the homeless, unemployed and the underemployed. Some of the most creative activities that people can perform are done in order to just survive in the neoliberal, globally competitive Creative City. Wilson and Keil (2008) offer the example of food banks in North America, where homeless people feign religious interest to obtain food from faith-based organisations – in effect they are acting. In Shanghai, I came across this abode shown in Figure 3.5, which had torn down a nearby promotional poster and used it as a roof.

However, being creative in order to survive (or to keep dry) is not a characteristic that aligns with the notion of the creative class and the Creative City to which it is so readily aligned.

Finally, running through these definitional critiques is the issue of the *the* in the creative class. To imply a unitary, singular, cohesive whole obfuscates the more heterogeneous, multiplicitous and diverse assemblage that is the creative class. There is a danger in collectivising such disparate groups of people and working

Figure 3.5 A makeshift roof in Shanghai.
Source: Author's photo, October 2012.

practices into a mobilised group for the reasons spelt out above; it glosses over political motivations, eradicates social tensions and, perhaps most critical of all, ontologically creates an 'outside', an 'other' – the 'uncreatives' perhaps. Be they the service, manufacturing, agricultural class or any other taxonomy, by valorising a certain group of people, it renders the others inconsequential to urban growth and vitality, simply by not being creative (in the economically tailored version of that word at any rate). Granted there is worth in identification and creating typologies (such as the one I have formulated to explain the critiques), but to do so without acknowledging the porosity and artificiality of the conceptually erected boundaries risks nullifying many critical processes and details.

On Tuesday, 25 October 2005, I travelled to the Manchester Convention Centre for the 'Enterprising and Creative Places' summit. A PhD student at the time, I was keen to hear the keynote by a certain Richard Florida. After a panel of speakers that included the late great Tony Wilson, Richard Florida stood up to deliver the closing keynote speech. He spoke of the importance of the creative class to the 'new knowledge economy' and highlighted how important the three Ts were in attracting these creative people to cities. The keynote concluded with a striking story from Seattle. He spoke of one of the Microsoft co-founders, Paul Allen, who built the Jimi Hendrix Experience Music Project (now the EMP museum) because of the inspiration that he took from Hendrix's music in founding Microsoft, one of the largest tech brands in the world. Suggesting that this valorised how a vibrant culture of bohemian lifestyle, tolerance and openness can bring technological jobs and growth, Florida's incidental vignette exemplified his creative class theories in a concise, striking, popular and 'cool' narrative.

However, even if we were to ignore all the definitional pitfalls articulated in the previous section and assume that there is such a coherent grouping that we can identify as potential growth-catalysers, the very mechanics of that process, the very calculative practices that imply that the creative class causes economic growth, have also been thrown into doubt. In other words, there is an issue with the *empirics*. The central empirical tenet of the creative class thesis is that the presence of creative people stimulates economic growth. This, however, has been questioned by a multitude of scholars, but was put most concisely by Malanga (2004: n.p.) who noted: 'The basic economics behind his ideas don't work.' The assault on the empirical backbone of the creative class thesis has been somewhat relentless, with the majority of the work focusing on how the data used to champion the creative class does not adequately account for the theory it supposedly supports. One of the major contentions first articulated by Glaeser (2004) and furthered by Markusen (2006: 1923) is that the 'regressions showing urban high-tech growth as a function of the presence of the creative class simply capture high human capital as measured by educational attainment'. In other words, the data captures those who are educated rather than creative (which points again to the ill-defined notion of creativity within the thesis). Even if there is a sense that the creative class could be considered anything more than simply those with a higher education degree, the main ideological thrust to the creative class's economic geography is that the jobs follow people – economic growth is a consequence of the presence of the creative class. However, this has been countered by a growing number of scholars including Storper and Scott (2009), who denote that while it may be partially the case that there is a group of 'creative' (or perhaps just educated?) people who move to certain cities because of the good mix of the 'three Ts', this must always be contextualised with a wider historiography of that city or region. They go on to argue that 'Most migrants ... are unlikely to be able to move in significant numbers from one location to another unless relevant employment opportunities are actually or potentially available' (Storper and Scott 2009: 161). In a study conducted in the US, Hoyman *et al.* (2009) are particularly damning:

> The creative class failed consistently across multiple statistical tests to explain either job growth, growth in wages, or absolute levels of wages. Additionally, the individual characteristics of the creative class – talent, technology, and tolerance – were negatively correlated with all our economic measurements. (Hoyman *et al.* 2009: 329)

The overarching critique then from an empirical perspective is that Florida has conflated correlation with causation – a common scientific problem no doubt, but one that nevertheless needs to be addressed and theoretically justified. By highlighting the concentration of creative workers in particular cities and claiming they are responsible for economic growth is a extremely large jump in logic, and one that, according to his critics, Florida does not explain adequately, if at all. In January 2013, Florida penned an article for *CityLab* (an online webzine for which he is the senior editor and which was formally known as *Atlantic Cities*) in which he argued that 'talent clustering provides little in the way of trickle-down benefits' (Florida 2013a: n.p.). This was seized upon by his critics, the most vocal of which (in the online arena at any rate) is Joel Kotkin, who argued that this admission proves that the creative class thesis, and its championing of 'hip and cool' cities is fundamentally flawed (Kotkin 2013). Florida's rebuttal was swift, arguing that he has always been 'aware of the inequality that is a by-product of the unvarnished creative knowledge economy' (Florida 2013b: n.p.). This tit-for-tat argument is indicative of the debates engulfing the creative class thesis, and how the empirical arguments are constantly challenged, questioned and then defended and refined.

Another well-versed and popularised critique was that put forward by Moretti (2012) in his book *The New Geography of Jobs*. In it, he outlines intricately the clustering of firms and the catalytic spurs that this can have on innovation. Drawing on Florida's analysis of Seattle (that he used so emphatically in his keynote address in Manchester), Moretti argues that Seattle is not the exemplar city of the creative class theory that Florida makes it out to be. In the 1980s it was 'gritty and depressing' (ibid.: 189) and was depopulating rapidly. He argues that Florida has got the formula the wrong way round, 'it became a lively cosmopolitan playground for educated professionals *after* it started attracting all the high-tech jobs' (ibid.), echoing earlier work by Storper (2010: 2034) who argued that 'skilled people appear in most cases to precede the creation of amenities, not principally to follow them.' Moretti also uses the example of Berlin stating that it is a cultural, vibrant city, a 'cool' place of edgy, post-communist aesthetics that speak to the kind of environment that the three Ts would epitomise. However, he says that 'there is only one problem with this picture: there are hardly any jobs' (Moretti 2012: 191). The city's high unemployment rate belies its 'creative vibe', which disproves Florida's theory. Again, Florida took to the webpages of *CityLab* to combat this attack. He claimed that Berlin 'is an extreme outlier; a very special case' (Florida 2012b: n.p.). Because of its problematic history many companies left the city and did not return when the Wall came down. He goes on to dismiss the entire 'which came first, the jobs or the people' argument: 'Given that we can't sort out causality, we shouldn't focus on just one side of the equation' (ibid.). The dismissal of causality is curious given the empirical effort that has gone into evidencing causality in his books. Whichever came first, the jobs or the people, and whether indeed causality matters or not, these are all red herrings when it comes to one of the fundamental issues with the empirics. Indeed many of the more general problems of the creative class thesis are repeated by Moretti's analysis. But given that the entire debate has it roots in American cities in the first instance, the theory is inherently centred on a Westernised, neoliberal, free-market approach to urban governance. It is an America-centric theory that has been retrofitted to other cities around the world. It is unreflective precisely because it is something that has been fundamentally designed not to analyse a city's economic ills, but to be adapted and *consumed* by other cities and regions all over the world. Thusly we move onto the *political* critique.

Such empirical critiques, while entirely justified, do not tell the wider story of how the creative class idea is actually having a deleterious effect on urban creativity rather than stimulating it. By engaging with Florida 'on his own terms' (as it were), by talking the same language of economic development and growth, is to not speak to the wider *political* issues that are inherent in using a creative class framework, i.e. one that adheres to a neoliberal ideology. This is because the applied political issues of the creative class are manifold, multiplicitous and ever-changing depending on the situation in which they are being discussed. Each city that the creative class is 'applied to' (or measured in) will have differing histories, cultures and social issues. As such, there are clear characteristics that render it extremely commensurable with a neoliberal paradigm (Peck 2005). Therefore, aligned with the discussion of neoliberalism in the previous chapter, we can borrow from assemblage theory (MacFarlane 2011) to articulate the 'emergent' qualities of the creative class, thus highlighting its ability to be adapted 'on the ground' by local governance structures (Springer 2010). As the creative class can be viewed in itself a part of the neoliberalisation of urban politics, there are a number of durable themes that engender inequalities with the urban realm which the following pages will isolate, unpack and critique. They are the brutality of statistics, glorification of precariousness and international mobility.

The continued justification of the creative class thesis is done so objectively and through the use of statistics, indexes, regressions and quantified variables, many of which have been countered with alternative sets of statistics (see Hoyman and Faricy 2009, for example). The papers and online articles that are produced to justify the creative class consistently use quantitative empirics to exemplify the main arguments, but in doing so depoliticises the central tenant of the work. In the same way that the language of urban growth uses words like 'redevelopment' and 'revitalisation' to render such processes apolitical, the use of quantified and statistical information performs a similar role. The very act of quantification is to transmogrify tacit subjectivity into an ordered objectivity (Porter 1995), to create functional order out of variegated and complex milieus. From a theoretical perspective, the very act of reducing the observed phenomena to statistical representation is to deny agency to those being studied. It is a reductionist process that strips away subjectivity and presents an objectivity that is calculable, and therefore malleable. It is, hence, a rather brutal, not to mention inherently political, act. Marcus Doel (2001: 555) puts it fairly succinctly when he states, 'I detest every one. No one in particular: just one in general. I prefer not to count on one. For me, number is a horror story. It is the most brutal of levelling devices.' Such brutality comes in the ways in which the perceived statistical objectivity obfuscates multiplicity. By offering up statistical representations of creative agency, the creative class thesis becomes functional in the mechanics of an urban entrepreneurial neoliberal system, a means by which current and future development plans can be justified. Moreover, Florida's work utilises city league tables that echo the listmania that the Global City paradigm sparked. Creative City index rankings litter Florida's work with the various indexes that he identified being coerced into a single 'creative city index'. Such hierarchical analysis further plays into political circles as being above other cities justifies the competitive practices employed. If a city can account for its financial investment by claiming that it directly causes a rise in the league tables, then this is a politically viable, and therefore much sought after strategy. By reducing creativity to an index, it removes any vagueness, ambiguity and subjective reasoning that are the antithesis of political economic strategising. There is of course a deeper politico-philosophical debate about the tyranny of statistics, one that cannot be fleshed out here. But in order to comprehend the political consequences of the formulation of the creative class thesis, the quantitative bedrock on which it is empirically founded is very important to its speed of uptake, international popularity and political longevity.

Moreover, the creative class thesis (in much the same way as the creative industries lexicon was posited) creates a situation whereby to oppose such an idiom is to be uncreative and all the stereotypes that are associated with that. Therefore, in order to sway politicians and urban leaders to allow developmental strategies, one only has to articulate it as 'creative' or attracting creative workers, as who would be able to argue against such a progressive line of reasoning? Peck (2005) articulates just how the 'creativity' language of urban policy has become completely undeniable. The creative class gains immense traction

... by the suggestive mobilization of creativity as a distinctly positive, nebulous-yet-attractive, apple-pie-like phenomenon: like its step-cousin flexibility, creativity pre-emptively disarms critics and opponents, whose resistance implicitly mobilizes creativity's antonymic others – rigidity, philistinism, narrow mindedness, intolerance, insensitivity, conservativism, not getting it. (Peck 2005: 765)

As Pratt (2008: 113) has similarly argued, 'who wants to be uncreative?' Indeed, this is perhaps the most endearing quality of the entire creativity paradigm from a neoliberal perspective, because it nullifies critiques from an anti-neoliberal body. The creative class context draws on work from Jane Jacobs, whose activism in New York in the 1950s and 1960s was fundamental in abating the Hausmann-esque re-zoning of New York by Robert Moses (Gratz 2010). Posthumously, Jacobs has been heralded as the 'champion' of diverse cities. Her polemic work on the multi-ethnic nature of New York's streets has been eulogised as saving the city from homogenous meta-planning. There is of course huge merit in Jacobs' work but in the same way that Florida is lamented for the unreflexive uptake of his work, Jacobs' ideas have been applied uncritically to today's modern cities as some sort of panacea for evils of large-scale urban development techniques. Her analysis concluded that it was the diversity, heterogeneity and complexity of cities that brought about economic growth through the constant combination, decomposition and recombination of ideas and hence creativity and innovation (Jacobs 1970). Such a view transposed to contemporary cities now contains gentrifying undertones, as the economic growth will inevitably create social inequalities and polarisation (Zukin 2010). Jacobs' work, however, is latterly seen untainted by such concerns and is championed when citing calls for more 'liveable' cities. The creative class theory (and the Creative City paradigm more broadly), by yoking to an ideology that is championing community values and street-level creativity, is blindsiding many critics. Those who oppose the financialisation of the city and large-scale, state-led investment attach themselves to the more utopian idyll of street-level development that is 'human-scale', focused on cultural difference and diversity and champions alternative ways of life. Who can refuse such idyllic urban characteristics? This is the genius of the creative class theory and its use by urban officials – it fetishises community-level development, but at the same time uses such aesthetic qualities to engender financial investment, real estate development and ultimately, gentrification.

Florida (2004) even uses the rather ugly and emotive term 'squelchers' (a term borrowed from Jane Jacobs) to denote those who are sceptical or attempt to block attempts at stimulating creativity, drawing further boundaries around those that bring prosperity and a creative utopia, and those that seek to resist it. Wilson and Keil (2008: 844) note this rather more harshly when they argue that the creative class 'flagrantly configures an elitist theme for change that feudal lords and bourgeois captains of industry in the past would have hesitated to do'. It seems that you are either for creativity or against it, there is no middle ground. But what this construction of elitism does is it creates a pedestal for precariousness. Innovation, creativity and flexibility also include precariousness, temporality and job insecurity, but these are not recognised. To be 'creative' within this framework is to be part of a workforce that, as has already been discussed, is increasingly character-ised by short-term and zero-hour contracts, unpaid internships, project-based labour and scarce freelance work. Of course there will be success stories, but this becomes problematic when it is articulated as the norm in lieu of a more meas-ured understanding of the vulnerability such normalisation can engender. As the creative class thesis continues to be adopted and used as a means to formulate urban and economic policy, then it is inevitable that such precariousness is woven into the fabric of the Creative City and we begin to see how creativity (and the glorification of the 'enterprising-self') is so amenable to neoliberal policies.

But perhaps the most neoliberal quality of the creative class are most acutely apparent in its international mobility. The uptake of the creative class thesis by urban governments first in the US, but since then across the world, has been immense. In the years after the publication in 2002 of the original text, Florida himself was booked for speaking events in a whole range of different cities (including the event in which I first heard him speak in Manchester in 2005). In one prominent example, Gibson and Klocker (2004: 430) describe Florida's visit to Sydney:

> At a major industry event, co-sponsored by *Men's Style* magazine, he was introduced by the MC as 'our chief inspirer', and was later joined by Sydney's planning elite in a five-star corporate luncheon ... Like a 'rock star' – Florida arrived, performed and left Sydney in a matter of days to continue the whirlwind Australian leg of his 'tROCC (the Rise of the Creative Class) tour.

The blatant branding of the creative class ideology as a 'tour' exemplifies its commoditised and internationally mobile state. Each city that booked Florida to speak, and even those that did not, were including his analysis in their planning documents, calling for investment from national governments and private inves-tors to buy into the creativity agenda. What was missing from this rhetoric, however, was a local reflexivity. Florida would of course pay tribute to the host city (as he did in Manchester with a warm embrace of Tony Wilson when walking on to the stage), but the central message would always be the same. Given that the message was about how that particular locality can prosper, it stands to reason that there should be an integration of the local histories, cultures, societies, econo-mies and idiosyncrasies. The heavily US-centric nature of Florida's work means that before enacting the policies that it entails, it should at least be broken down and tested within a local context. Gibson and Klocker (2004: 432) go on to say:

> Such visitors [Florida] continue to be accorded favourable 'star' status and shielded from criticism, even though they are unlikely to know much about local circumstances. Richard Florida was even asked by the New South Wales Deputy Premier Andrew Refshauge, at a corporate luncheon in Sydney, how to 'creatively' deal with poverty and crime afflicting the city's 'problem' Aboriginal community.

This exemplifies just how appealing the creative class theory can be to urban leaders who see it as a panacea for their own local issues and 'problems'. Peck (2005) has highlighted how the creativity agenda filled a vacuum of urban entrepreneurial strategy: 'Whatever else it may be, Florida's creative-city thesis is perfectly framed for this competitive landscape, across which it has travelled at alarming speed' (Peck 2005: 767). The 'alarming speed' of the uptake of the thesis across the cities of the world has continued in the decade since Peck's paper and speaks to the way in which the creative class theory fitted so neatly with the prevailing urban entrepreneurial infrastructure of the twenty-first century. The timing of the thesis was perfect, the message was undeniable, the empirical backbone was clean and concise. As Peck (2005: 764) has said, 'both the script and the nascent practices of urban creativity are peculiarly well suited to entrepreneurialized and neoliberalized urban landscapes.'

Florida himself, since the rise (and fall) of the creative class theory has been forced to take to numerous platforms (most vehemently, though, online through *CityLab*) to defend his work, but also to decry the uptake of the idea by various cities. He has quite rightly pointed to the fact that urban governance has cherry-picked from the creative class theory to justify whatever project they are pursuing at the time. Like the Creative City Policy document from the City of Sydney discussed in the previous chapter, the creativity paradigm can be applied to just about any infrastructure project, education programme, event, place-based beautification strategy or business-incentivisation policy. The creative class theory has become something of a feral beast, being used by urban officials, private companies and community leaders as a linguistic toolkit to extol the 'creative' virtues of their particular agenda. And therein lies the crux of the matter – the creative class provided the final piece in the jigsaw of neoliberal urbanism of the twenty-first century. The now entrepreneurial urban governance structures that were using the creative industries as an industrial conduit to urbanise and industrialise culture had a functional, defendable and internationally recognised policy vehicle with which to operationalise the spread of neoliberal policies. The Creative City could now take shape. But what is it to *physically* look like? What buildings would be needed? What about housing? The physical infrastructure of the Creative City is absolutely fundamental to the neoliberal institutions as it is where the overwhelming majority of profit is to be made. In this respect, the work of Florida too has great influence. The creative class is critical to prosperity – of this there was no doubt. But to attract them, the theory goes that a city needs certain amenities, certain 'creative' infrastructures in which the creative class as well as the creative industries could proliferate. They need places to work, places to live, places to eat, drink and play. It is these physical places, and their imbued social, cultural and economic problems, that are the subjects of the next chapter.

4 Quartering creativity

Real estate in the Creative City

The initiation of the creative class and with it the realisation of the Creative City and its viral-like spread around the cities of the world has had nothing short of a seismic shift on urban governance in the twenty-first century. The perceived prosperity and economic growth that the creative class bring to a city has ignited a wave of projects, policies and processes that have been enacted by city governments designed to 'attract' these creative people to their particular city. The 'soft' branding techniques outlined in Chapters 2 and 3 form part of a broader remit of policies aiming to formulate the Creative City. These initiatives, given that they are specifically designed to attract creative people and creative industry firms and businesses, therefore have a *systemic* need for physical urban space. All these creative people are going to need somewhere to live, to work and to consume. These spaces not only need the practical and infrastructural attributes required by creative people and companies but also the aestheticised form worthy of commensurability with the 'Creative City' brand and the political economic spectacle therein imbued. As such, the physical infrastructural development that takes place under the rubric of the Creative City paradigm in many ways is the natural and logical successional step; indeed it is the manifestation of the 'spatial control' necessitated by neoliberal urbanisation (Lefebvre 1991). The realities though are obviously far more complex and the causal relationship between a Creative City brand and the creation of physical Creative City space is not a linear and unidirectional one. The Creative City paradigm includes both the 'soft' marketing strategies and the 'hard' infrastructural urban development in tandem and in parallel, often feeding back into one another in a continual cycle of Creative City urban governance agenda-making. Whether it is the construction of a brand-new 'Media City', the reclaiming of derelict and disused industrial land to build a 'Cultural Quarter', the reigniting of a halted development under more 'creative' branding or (what tends to be the case) a hybridised mix of all three of these, the making-physical of the Creative City is an integral part of the policy as it generates the most *wealth*. As such, it is part of the broader urbanisation trend noted seminally by Lefebvre (1991, 2003 [1970]), which entails the increasing control and development of urban space as part of a political desire to control state mechanisms. So it is of no surprise that there are large real estate companies with a vested interest in 'creative' zones, prioritising the physical built form, most readily through the rent gap theory (Smith 1987), as a way of maximising economic benefit from the now renewed policy drive of attracting creativity (in all its economic guises). Moreover, the increasing intermeshing of these private and corporate power relations with the local and city councils further exemplifies the entrepreneurialisation and corporatisation of urban spatial governance, as outlined in Chapter 2. As such, the zoning of creativity in the city is a critical part of the Creative City paradigm as it creates the linearised, neatly packaged narrative that can be easily marketing under the related branding procedures.

These creative zones then are paramount to the 'success' of the Creative City in terms of ossifying the brand and creating the attributes needed to climb the city league tables. This chapter then will outline their continued justification, construction, aesthetics and management as part of the drive to facilitate the political economy of the Creative City paradigm. Moreover, the pressure for cities to find the next 'quick-fix' so as to not fall behind in the global race to be creative is creating a market for 'off-the-shelf' creative zones. Therefore in many cases (particularly in the UK and more widely in the Global North, and increasingly in the Global South) real estate companies and other urban planning service providers (architects, creative consultants, interior designers and so on) collaborate to create models that can be presented and sold to urban governments desperate to compete in the new landscape of creativity. These creative zones can take many varied forms, and it would be inconsistent with the ever-evolving, imminent nature of neoliberal Creative City assemblage to suggest that there is a 'static' typology. However, the existence of certain 'models' of creative zones is very much imbued within the instrumentalisation of Creative City policy (not to mention the repeatability of policies that is symptomatic of neoliberalism) – the want to be 'creative' is often counterbalanced by the necessity to achieve this on the cheap. And this is only exacerbated by the post-2008 current climate of austerity, which brings with it a diminution of public funds and political will toward 'culture' in lieu of more 'fundamental' public services such as health, education, transport and welfare. Also, the desire for the short-term fix, often catalysed by the nature of political cycles, further serves to quicken the uptake of these urban templates and models. So when private investors offer the chance to create one of these 'creative zones' quickly and at a reduced cost to the public purse, they are taken up without criticality and with the minimal amount of public consultation that the constitutions will allow, and the associated instruments of the privatisation of space are imprinted upon that area, often to the detriment of the creative milieu that was sought after in the first instance.

This process then leads to the 'serial replication' (McCarthy 2005) of a particular type of creative zone or district. In this chapter, I want to focus on two instances of these 'creative zones', the Cultural Quarter and the Media City. These two represent perhaps the most 'identifiable' model of the physicality of the Creative City, in that there are mechanisms of commercial and corporate networks whose sole purpose is to replicate and 'sell' these ideas to urban governments. Another justification for the increased scrutiny is that they have regional and geographical linguistic specificities. In other words, the Cultural Quarter terminology is used predominantly in the UK and Western Europe whereas the Media City lexicon (for reasons that will become apparent through this chapter) is being utilised in cities in the Middle East, Asia and Africa as well as in Europe. In the UK, for example, since the 1980s there has been a rapid uptake of the so-called Cultural Quarter. Today, there are scores of cities and towns throughout the country that have built or are in the process of building a Cultural Quarter, all with similarities in terms of architectural form, branding and/or economic activities catered for (mostly cultural consumption). Therefore with the homogenisation of a city or town's cultural offerings comes the actual reduction in individuality or uniqueness of the area, and hence is fundamentally counter-intuitive to the promotion of creativity and innovation. Coupled with formulaic and generic branding features, these Cultural Quarters are simply cata-lysing the homogeneity of the UK's townscapes, and as a result are becoming another neoliberal practice that catalyses a 'UK' urban aesthetic. 'Media Cities', however, are examples of these creative zones that exist across the globe. These large-scale, expensive and highly corporatised areas represent the latest incarna-tion of the 'high-tech fantasies' (Massey *et al.* 1992) that were the go-to business park developmental idioms of the late 1980s, and are characterised by the priva-tised, securitised and consumption-orientated characteristics that have become synonymous with the contemporary city. The global competitiveness engendered by the Global City paradigm causes those cities aiming for the 'top' (i.e. those in the 'second tier' of Global City status – the GaWC Beta cities) to look for that quick policy initiative that will propel them up the hierarchy, and the Media City model fits that bill. As such, the Cultural Quarter and the Media City are increas-ingly popular as they operationalise the greatest amount of profit-generation of the Creative City paradigm. By no means though are they the only form of crea-tive zoning practices. There is a myriad of terms – creative clusters, cultural districts, media quarters – and they all espouse a similar urban aesthetic. Whatever the geometric and topological form though, they all have similar constitutive, mechanistic, maintenance and aesthetic qualities to those of a Cultural Quarter or a Media City. It is to each of them that I now turn.

Cultural quarters

Leicester is officially the tenth largest city in the UK, with a population of around 350,000 people, of which I was one for some of the most important and formative years of my life. The city was the hosiery, shoe and textile centre of the UK in the nineteenth and early twentieth centuries. However, like many other cities around the country, it suffered from deindustrialisation in the postwar period. And more recently, it has also suffered from city centre decline in terms of lack of footfall, migration of businesses to out-of-town retail parks and an over-reliance on a night-time economy and all the antisocial behaviour that comes with it – the same old story that blights many UK regional towns and cities. The city council, however, was keen to redress this decline and initiated a number of schemes including the complete overhaul of the city centre shopping centre and the creation of a 'Cultural Quarter' in the St George's area of the city. The area is adjacent to the city centre and was home to a number of piecemeal businesses, surface car parks, a large office supplies warehouse, an industrial printing house, a nightclub, a 'swinger's club' and numerous car mechanics. As such, it was the ideal location for redevelopment given its geographical location close to the recently 'revamped' city centre, promiscuous clientele, low ground rent and lack of a coherent 'identity'. In 1999, the city council embarked upon an ambitious project to brand the area Leicester's Cultural Quarter. As of 2013, the council claimed to have 'transformed the St. George's area of Leicester from the city's former textile and shoes hub into a thriving area for creative industries, artists, designers and crafts people' (Visit Leicester 2013: n.p.). It now includes creative business incubators (one called the Phoenix Centre – immediately conjuring up an imagery of 'rebirth'), cinemas, cafes and bars, but the jewel in the crown is 'Curve' (see Figure 4.1). In 2008, Curve opened having cost £61 million, nearly double the original quote of £35 million. In 2009, the Audit Commission stated that there was 'weak project management' (BBC 2009: n.p.) and that 'Curve did not manage to control significant increases in cost' (ibid.). Despite Curve coming in significantly over budget, the leaders of the Cultural Quarter have argued the

Figure 4.1 Curve (on the left), Leicester's CQ.
Source: Author's photo, July 2013.

local area of St George's has benefited from like-for-like private investment since Curve was commissioned:

> We have accounted for approximately £60 million worth of private sector investment in the area … these buildings didn't start to develop until we started to lay the foundations for Curve. (Candler, quoted in BBC 2009: n.p.)

However, given more recent developments in the wake of the financial crisis, many auxiliary leisure developments such as restaurants and bars have closed and been repossessed by landlords, with the business owners citing low footfall. Residency in the refurbished flats is low (Figure 4.2) despite new apartment complexes being planned (Figure 4.3).

Also, Curve received a £1.03 million 'Sustain Fund' from the Arts Council in 2010, a fund dedicated to help art institutions that are financially struggling because of recessionary pressures. And Curve is not the only institution within the Cultural Quarter to suffer. Phoenix Square, which includes the Digital Media Centre, has seven offices designed for short-term creative industry business usage (i.e. incubator spaces). Located adjacent to Curve, the Phoenix development is a key proponent of Leicester's Cultural Quarter given its mixed use of creative

Figure 4.2 Rental boards in Leicester's CQ.
Source: Author's photo, July 2013.

Figure 4.3 Advertisement for residential construction/renovation projects in Leicester's
    CQ (on the left) affixed to a scaffolding company building.
Source: Author's photo, July 2013.

industry production and consumption space. However, in 2010, it received a
financial boost of £250,000 from the local council after managers admitted to
local news outlets it was losing cash substantially. To date then, Leicester's
Cultural Quarter is not without its problems, but the erstwhile conduct of the city
council in developing, maintaining and pushing through the idea in the face of
large financial pressures indicates the importance of the scheme to the city,
clearly ossified by the council's (ultimately unsuccessful) bid for the 2017 UK
City of Culture, which only adds to this policy drive (although the whole City of
Culture scheme is another area of debate – see Cox and O'Brien (2012)).

Therefore this drive for urban 'culture-led' regeneration is mirrored in Cultural
Quarters across the UK, Europe and, increasingly, the world. Another example
can be seen in Shanghai, where there has been a concerted governmental push to
implement 'creative industry clusters' across the city. Zheng (2011) highlights
how there are no less than 75 of these parks across the city (of which I managed
to only see only a fraction). One of the most prominent, M50, bares striking
aesthetic characteristics with other cultural quarters and creative 'zones' in the
UK and European cities. Whitewashed walls, reused industrial spaces, boutique
art galleries, outdoor public art (see Figure 5.4) – they are all internationally

Figure 4.4 M50 creative park, Shanghai.
Source: Author's photos, 14 October 2012.

Figure 4.5 The plaque for M50, Shanghai.
Source: Author's photo, 14 October 2012.

recognised characteristics of a cultural 'zone'. Also, Figure 5.5 shows the placard upon entering M50 and how it is officially part of the 'Shanghai Economic Commission'.

Shanghai's deliberate policy of building more of these 'zones' exemplifies the increased global reach of the Creative City paradigm, and moreover its commensurability with urban entrepreneurialism. As Zheng (2011: 3577) notes: 'As a cultural approach towards city entrepreneurialism, [Shanghai's Creative Zones] embody a rising trend that combines flagship leisure venues and upper- and middle-class urban lifestyles, with a flavour of urban cultural heritage and cultural tourism' – in other words a direct replication of the Cultural Quarter policies that have been economically successful (or not) elsewhere.

However, before continuing with the examination of the ways in which these developments are embroiled within a Creative City politics, and therefore impacting upon the inherent creativity and cultural fabric of the city, it is perhaps pertinent now to ask a simple yet often unsatisfactorily answered question: what exactly is a Cultural Quarter? There is an extensive literature on the prominence of Cultural Quarters (particularly in the UK) with varying (and sometimes ambiguous) definitions. Cultural Quarter definitional literature has a historical determinism, one that is based on the foundational desire of urban governance to 'zone' cities (to create identifiable 'quarters'). However, the origin of the term and the urban 'revitalisation' it represents is coupled to the development of locally based cultural industry policies in a few UK cities during the 1980s, namely Sheffield, Manchester and London (and later, of course, Leicester). Brown *et al.* (2000) have noted that the deindustrialisation that was rampant in UK cities (particularly in the North and as has been noted in Leicester also) in the 1970s and 1980s offered many areas in and around cities that are prime for redevelopment. Moreover, and as part of the cultural turn in urban entrepreneurial activity (outlined in Chapter 2), these areas required minimal retrofitting, given that the large industrial, cavernous rooms lend themselves easily to exhibition and gallery spaces (see Figure 4.4) or for cultural workers who require large spaces at low rental costs. Since the 1980s though, as was also outlined in Chapter 2, the shift in political language from culture to creativity brought about by the DCMS definition of the 'Creative Industries' in 1998, the term Cultural Quarter has been tweaked, adapted and/or sexed up resulting in the proliferation of the different terms detailed above. We therefore see 'creative quarters', 'creative hubs', 'cultural clusters' and other variations on the theme (Oakley 2004; Evans 2009), but they all have the same 'feel' and similar etymologies (even when transposed to Shanghai). The first word will be the adjective that describes the purpose of the area and the expected practices therein. Creative, cultural, media, museum, heritage – or to be more distinctive, a term that links it to the local histories may be used such as Belfast's or Derby's Cathedral Quarter (centred as they are in the areas around their cathedrals), or Belfast's other 'quarter' the Titanic Quarter, named so as it was where the ocean liner the *Titanic* was built. The second word of the phrase will be the spatial aspect – quarter, cluster, zone, park, street, etc., depending on the size of the area or the constitution (mainly production or consumption activities). Also, just one (mix-used) building can be constructed or retrofitted and branded in such a way as to label it as a 'quarter' or cultural centre of some sort despite not including the 'open', publicly accessible areas that are often associated with Creative City-specific zones. But whatever their constitution, marketing them in a way that buys into the prevailing creativity paradigm is essential. This is of course linked to the branding strategies that are so critical to urban governance strategies and the whole need for the lexicon of urbanism to reflect a commercial and unproblematic utopia (Peck 2005; Kornberger 2012). What has also been important is the 'Porter effect', namely the near ubiquitous prevalence of the 'cluster' in government economic policies at a national and urban level (based upon the seminal economic geographical work of Michael Porter). This has seen the preponderance of the term 'cluster' creep into the vernacular of cultural development, as it alludes to the economic benefits of zoning, without foregrounding the implicit problems of such a strategy. At any rate, with these 'augmentations' of the 'Cultural Quarter' as a definable term (from the 1980s), they have used the contemporary political economy narrative to couple cultural-led development to the economic 'regeneration' of cities. Be it through using the language of 'creativity' (which brings with it perceived innovation and entrepreneurial activity) or 'clustering' (which brings with it perceived wealth generation by simply 'being there'), Cultural Quarter language has mutated over time. Other research and analysis of Cultural Quarters have defined them through the type of cultural provisions they offer, be they museum, retail, digital or media based (Santagata 2002) or whether they are production or consumption-orientated (Pratt 2004). Despite the increased variation, as we shall see, they remain fundamentally a strategy of capitalist urbanisation. The 'perfect storm' of the availability of large areas of the city through deindustrialisation, the ease and low cost of redevelopment, the importance of culture and creativity to economic growth (as part of the new urban political mantra) and the need for a short-term financial boost meant that these Cultural Quarters have become the go-to policy model for urban 'regeneration'. To date, there are over 35 initiatives in the UK that can be characterised as a Cultural Quarter scheme. Indeed, as Oakley (2004: 68) noted: 'No region of the country, whatever its industrial base, human capital stock, scale or history is safe from the need for a ... "cultural quarter".'

So if a city has identified a Cultural Quarter as the cure to their current urban blight, how does it go about building one? As we have established, public money is sparse so a long-term consultation process or public inquiry is viewed as an inefficient use of funds. However, there is a myriad of consultancy and private advocacy institutions ready to offer urban councils insight and evaluative capital on the best way to construct, manage and maintain a Cultural Quarter. Often operating on the cusp of many different disciplines including town planning, architecture, interior design, management consultancy (and even academia), these consultants offer descriptive and rather unproblematic appraisals of how to best design a Cultural Quarter. One such account is from Montgomery (2003, 2008), who outlines an intricate prescription for a 'successful' Cultural Quarter. Among a nuanced list of consumption and creative and cultural industry production facilities, he argues that:

> An essential pre-requisite for a cultural quarter is the presence of cultural activity, and, where possible, this should include cultural production (making objects, goods, products, and providing services) as well as cultural consumption (people going to shows, visiting venues and galleries). (Montgomery 2003: 296)

He is keen to stress the importance of 'both sides' of the cultural and creative economy rhetorical binary, production and consumption, to the success of a Cultural Quarter. The former creates wealth and profit (through the clearly outlined benefits of the clustering process) and the latter is the means by which such wealth and cultural benefit can proliferate throughout the local and subsequently wider urban community. He goes on to suggest that there is a triumvirate of characteristics that are fundamental to a Cultural Quarter. They are *activity*, *built form* and *meaning*. Activity covers a range of cultural and creative economic activity from the 'strength of [a] small firm economy' to the 'presence of an evening economy' (Montgomery 2008: 309). Here, the preponderance of the economic determinism is already beginning to be exposed and a lacuna of cultural and creative activity that does not have direct (or indirect) economic functionality is becoming acutely obvious. Second, the built form of the Cultural Quarter must contain 'fine grained urban morphology' and 'an amount and quality of public space' (ibid.). How much 'quality public space' is needed is not elucidated, but in analysing the built form, he goes on to suggest that:

> In the more successful quarters this design ethos is carried through into architecture (modern, but contextual in that it sits within a street pattern), interior design (zinc, blonde wood, brushed steel, white wall) and even the lighting of important streets and spaces (ambient, architectural and signature lighting, as well as functional). All of these reinforce a place's identity as modern and innovative. (Montgomery 2008: 307–8)

Essentially, then, there is a suggestion that to be innovative, a CQ has to have very specific architectural styles and use particular materials that represent the contemporary working environments of the modern and knowledge-based economy. Finally, the Cultural Quarter has to have meaning, which Montgomery (2008: 310, original emphasis) argues centres around culture, as 'culture after all *is* meaning'. He also goes on to point out that authenticity is crucial as is the ability to innovate and change.

This prescription of a 'successful' Cultural Quarter then is highly nuanced, detailed even to the interior finish and materials to be used. However, it is clear from his account that it is one based on the 'updating' of activity to be more in line with those conducive to late capitalist economic patterns (Pratt 2008). Economic activity such as the knowledge economy, innovation, the creative industries, the experience economy, urban place-based festivals and cultural events are those that are the apparent panacea for economic stagnation and decline. Within a short-term timeframe, it is clear to see why such an account is attractive to urban councils; it ticks all the boxes of contemporary urban and economic development. However, by prescribing a step-by-step guide as to how an urban area can be developed into a Cultural Quarter is to neglect the importance of the variance of place, as well as to neglect urban creativity *in toto* in favour of a purely financialised pastiche.

Specifically, then, let us turn to each of the terms in Montgomery's triumvirate of concepts that are required. In reference to *activity*, Montgomery (2003, 2008) outlines the range of economic factors needed for a Cultural Quarter to be 'successful' (presumably here success is defined along economic and wealth-generation lines). However, in failing to take into account the creative forces of urban actors already in situ by having a limited (or purely financial) view of creativity, there is a real danger of reducing a Cultural Quarter development into a social gentrification project. Montgomery is keen to point out that the development should already be built upon existing cultural and creative strengths; however, if the existing 'creative' strength does not adhere to (or have the potential to be remoulded into) those activities clearly defined as critical success factors, what is to be done with them? Will they be able to afford the new rental costs? Will they have the right image that is in line with the expensively assembled brand? Many instances of Cultural Quarter development in the UK have seen existing firms bought out or given compulsory purchase orders (such as in Leicester), homeless people and squatters evicted from derelict properties (Clement 2012) and urban subcultural activity illegalised and displaced. A contemporary and highly relevant example is the South Bank in London, the self-proclaimed cultural quarter of London. The development of the South Bank Centre into the 'Festival Wing' announced in May 2013 included the removal of the skating area in the undercroft part of the structure, with retail outlets to be built in its place. This process and the subsequent campaign that saved the skating area is detailed later (in Chapter 7); however, it is a clear and present indication of the blatant disregard these Cultural Quarter developments have for activities that do not or cannot be directly profited from via formalised commercial activity.

In terms of the second of the conceptual trinity, *built form*, Montgomery highlights how there is a desire to change the allocated urban area into a 'modern and innovative' urban space by having very specific building materials (zinc, brushed steel, etc.) and appropriate street lighting creating a very specific commercial and 'innovative' aesthetic. The specific design of buildings within a Cultural Quarter is of course not a derivative from the local urban council, but the architects, construction companies and interior design firms are part variegated systems of processural space-mobilising constructions (Wilson 2004). The networks of private companies in contractual negotiation with the commissioning councils (often with asymmetrical power relations) will pinpoint the built form that is seen to be conducive to cultural and creative industry production. But in doing so homogenised office provisions and replicated 'incubator' spaces (Turok 2003) proliferate, narrowing the resource base for those cultural activities that do not conform or require these very specific (and often expensive, in terms of rental costs) urban spaces.

Finally, in reference to *meaning*, Montgomery (2008: 310) suggests that a 'cultural quarter which produces no new meaning – in the form of new work, ideas and concepts – is all the more likely to be a pastiche of other places in other times'. He then goes on to argue, 'a good cultural quarter, then, will be authentic' (ibid.). Notwithstanding the inherent problems of attempting to 'replicate' authenticity within contemporary urban renewal (see Zukin 2010), by suggesting that meaning is only produced through new work he is reducing the cultural and social sensibilities of place to mere unconnected non-sequiturs. A lack of (economically) manufactured meaning, Montgomery (2008: 310) suggests, creates Cultural Quarters that are 'simply a collection of publicly funded venues ... or else an emblem of former culture – "heritage"'. This is to suggest then that a place devoid of economic action and consumption is a place devoid of meaning. The suggestion that publicly funded venues and activities do not produce new meaning is, again, highly instrumental and highlights the commercialism that is rampant in the culture-led development of the urban. Again, this reductionist approach negates and indeed defenestrates peri- or non-economic activity as spatially motivated activities. The diminution of heritage to mere emblematic discourse belies the cultural influence such areas can exert, yet because of their inability to be directly accountable financially, heritage functions are seen as part of the 'old' interfering with the search for the 'new'. However, the celebration of history is a key facet in the meaning of a Cultural Quarter and not a reduction of it.

Another very common prevailing characteristic of a Cultural Quarter is the presence of a 'flagship institution' (Evans 2009; Mommaas 2004; Pratt 2008). The concept of a flagship institution to stimulate economic growth is a traditional urban renewal strategy (Smyth 2005), with Bilbao's Guggenheim Museum often cited as the exemplar case of how a cultural institution can kick-start an ailing economy, with the process even entering urban strategy parlance as the 'Guggenheim effect' (Plaza *et al.* 2009). The economic success of this venture has not gone unnoticed around the world. As such, countless attempts have been made by cities to build a one-off cultural institution (be it a museum, art gallery, theatre, performing arts centre or other 'high culture' asset) in the hope that 'trickle-down' economics will take hold and catalyse economic activity in the form of tourism, consumption and linked production. Moreover, often these institutions will be designed by internationally renowned architects (or 'starchitects' (Knox 2012)), who are offered increasingly large amounts of money to design brash and 'iconic' structures that will help that particular city stand out. However, the perceived uniqueness offered by such outlandish architecture is countered by the increasing number of such developments globally. Also, as has been noted in countless studies from many places around the world (including California, USA (Grodach 2010), Oslo, Norway (Smith and von Krogh Strand 2011) and East Asia (Kong 2007)), the variance in 'success' is massive, and is often linked to a variety of contextual factors that are beyond the control of the instigators of the initial project. For example, as Smith and von Krogh Strand (2011) note, the Opera House in Oslo succeeded through having local and national ambitions, not an international or global outreach from the start (these came as unexpected consequences). Also, Grodach (2010) found that in California, external factors such as the composition of and local demand from arts establishments were key, something that cannot be engineered so easily via traditional flagship strategies. Despite numerous warnings and cautionary tales of the need for local reflexivity, many cities have utilised the flagship institution strategy to instigate the creation of a Cultural Quarter. Already in this chapter we have seen how Curve in Leicester was built at huge cost (the architect of which is the international recognised Rafael Viñoly) and currently is struggling to maintain projected income levels. Other examples include the catastrophic failure of the National Centre for Popular Music (NCPM) in Sheffield, UK (again, a building with a striking and 'unique' design). Put forward as a millennium project, construction was completed in 1998 and it officially opened in March 1999. Sheffield council commissioned the NCPM, drawing on lottery funding in an attempt 'reassert the local within global cultural flows' (Brown *et al.* 2000: 440). Costing £15 million to build, the NCPM hoped to attract half a million visitors per year according to the promotional material. With poor visitor numbers and a failed £2 million re-launch as a live music venue, it was bought by Sheffield Hallam University in 2003 and is now their Student Union building. The positioning of the NCPM in Sheffield's Cultural Quarter was a deliberate ploy to stimulate the local creative industry community and complement the vibrant music industry within Sheffield (Brown *et al.* 2000). However, despite the relative success of the surrounding incubator spaces and vibrancy created by the adjacent Sheffield Hallam University, the low level of visitor numbers was not enough to maintain the NCPM economically, and hence it had no other option but to shut down and be sold off.

Therefore it is clear that currently there is a definitive 'model' for Cultural Quarter production, specifically in the UK. Indeed, with the analysis of Montgomery's (2008) narrative above, there exists a 'model of best practice' that is adopted, largely uncritically, by town and city councils. Landry (2006) and the late Simon Roodhouse (2006) have also produced processional 'how-to' guides on constructing Cultural Quarters, further adding to the policy-orientated litera-ture that is readily available (and all too accessible) to urban councils looking for the quick fix of urban 'renewal'. Such a Cultural Quarter model then has been refined over the years. The rather clunky policy processes that blighted the 'first' Cultural Quarter in the UK, Sheffield (along with its monumentally unsuccessful flagship scheme), have been slowly massaged through a honing of consultancy language, a better understanding of market forces and the slow erosion of public planning and bureaucratic mechanisms into a well-oiled and packaged policy product. Given the way in which the Creative City paradigm has been predicated upon a systemic desire for efficiency in urban development and a replicable model of short-term commercial urbanisation, the Cultural Quarter model very much fits with such a narrative. Indeed, they mutually constitute each other and stem from the same ideological kernel. Cultural Quarters therefore are simply a mechanistic variance in the wider drive of the creativity paradigm, an alternative language of the Creative City that fits the local and national political governance. These models, as they continue to be replicated across cities, only serve to decrease individuality, creating homogenous cityscapes of consumerist, econom-ically deterministic cultural provisioning. However, the private investment and real estate firms that provide such a homogeneous aesthetic benefit massively from such a model. Despite spectacular examples to the contrary (such as Curve), these are the exception rather than the rule and the costs of Cultural Quarter production continue to drop via efficient refinement or design. The income stead-ily rises via incremental rent increases and the sheer number of Cultural Quarters being built signals multiplication of these profits. Therefore the model is a self-serving and financially lucrative one, so the interest is to maintain it. However, if Cultural Quarters represent perhaps a model of the Creative City that is predomi-nantly UK-based, Media Cities, then, are the global equivalent.

Media Cities

The construction of Media Cities is gaining in popularity across the various regions of the world. While the term 'media city' is relatively new (and perhaps slightly popularist), they can be defined as large, planned, highly developed urban areas designated specifically to *concentrate* media and creative industry production in it's broadest sense (Mould 2014a). As such, they continue the political motivation to cluster – Silicon Valley is still idealised by urbanists and politicians as the clustering model par excellence and processes are in place that are constantly trying to replicate its success. Moreover, media cities are designed in many ways to be self-sufficient. Unlike Cultural Quarters or stand-alone physical investments in line with a Creative City paradigm that (have an intention at least) to integrate with the city, Media Cities act as islands of investment, citadels of 'creativity' that are self-serving and stand alone. A number of examples of Media Cities across the world can be cited, the most recognisable being Dubai Media City, DR Byen in Copenhagen, Digital Media City in Seoul and MediaCityUK in Salford. Other similar developments include the Australian Technology Park in Sydney which, while not being branded officially as a Media City, there are multiple references to it within industrial and governmental policy literature, as well as blogs, magazines and general industry 'chatter' as 'Sydney's Media City'. Some examples have also branded themselves uniquely, with Abu Dhabi's Media Free Zone called TwoFour54 (a reference to the development's latitude and longitude coordinates). However, these developments will more often that not have large office, studio and exhibition spaces (usually at high rental costs), and house auxiliary leisure and cultural services. Goldsmith and O'Regan (2003: 33) have noted that the Media City has been 'recast as a form of commercial property/industrial park development' which has a primary goal of attracting international business, and so in many respects they can be viewed as the contemporary iteration of what Massey *et al.* (1992) called 'high-tech fantasies'. As will be seen, they are like the Creative City paradigmatic utopia in microcosm, enclaves of creative class professionals living, working and consuming in readily available units, privatised, corporatised, heavily branded, panoptic pastiches of urban creative activity.

What sets Media Cities apart from other hard-infrastructural Creative City initiatives (such as Cultural Quarters, creative zones, incubator spaces and so on) is the gargantuan level of financial investment needed. The technological capacity, luxury condominiums, hotels and high-end office provision needed is far beyond the price range of city governments (with, however, notable exceptions, particularly in the Middle East), and as such the levels of sunk financial investment to construct these Media Cites is vast. Moreover, because of this astronomical level of investment and the level of finance therefore expected, they are providing a highly corporatised industrial environment that is characterised by state-of-the art technological provision and luxury real estate, all at an extremely high cost. In combination, these areas are providing the whole Creative City 'package' – the high-specification working facilities, the luxury living accommodation, the high-end cultural and retail outlets and the secure (often gated) location. These areas are one-stop shops for the creative class.

In terms of the technology, Media Cities have super fast broadband connectivity in order to transfer the large (super high-definition quality) content, state-of-the-art production and exhibition facilities and global connectivity (Evans 2009). MediaCityUK in Salford (in Greater Manchester, UK), was financed by Peel Holdings (at a cost of a reported £1bn) and boasts broadband speeds of 10 giga- bytes, which is needed by many of its tenants, including the BBC (the British Broadcast Corporation, the UK's national media institution) who have to transfer high-definition film between Salford and the corporation's other major base in London. It also has high-end TV studios, editing suites and radio studios of contemporary usage all requiring high-definition (and in some cases 3D) capabili- ties. All of these capabilities were financed and owned by Peel Holdings, the owners and landlord of MediaCityUK. They rent the facilities out to the tenants (including the BBC), thereby increasing the cost of usage and the reliance of the tenants on the landlords. Copenhagen's Media City (known as DR Byen, DR being the Danish national broadcaster and Byen being Danish for town: 'DR Town') also contains such high-technological capacity, something that was outlined in the initial planning: 'which must offer the optimal flexibility in terms of fitting-out, options to realise new production and working practices and have room for an organisational change of editorial staff and groups to suit changing programme outputs etc.' (Jensen 2007: 21–2). Another feature of such areas is the mediation of the outdoor spaces, with large video screens and walls being a common site. Figure 4.6, taken at Digital Media City in Seoul, shows how the side of one of the

Figure 4.6 Video wall in Digital Media City, Seoul.
Source: Author's photo, October 2012.

recently built office buildings has a large video screen embedded within it. Reflected in the screen, the continual development of the area is seen, as indicated by the unfinished skyscraper with a construction crane atop.

Figure 4.7 shows a smaller information screen in the Australian Technology Park in Sydney. Smaller, but more frequent, these interactive touch screens give visitor and tourist information about the Park. This digitisation of the outdoor spaces not only allows for more sophisticated forms of outdoor advertising (further commercialising the topology of these spaces), but represent what Paul Virilio (1994) saw as 'media buildings', structures that convey information rather than habitation. The removal of any indication of actual community living in lieu of a branded, slick, digitised and commercialised environment is commonplace within these Media Cities.

There are of course exceptions to this technological characteristic, notably within Dubai Media City, which is an economic 'free zone' within Dubai. These free zones are agglomerations of office buildings designated by the Dubai city municipality to be exclusively dealing in one particular industry (Bagaeen 2007). So there are oil and gas free zones, jewellery free zones and, in the case of Dubai Media City, a media free zone. There is little in the way of extra industry-specific provisioning in these zones, and so initially Dubai Media City had no extra

Figure 4.7 Video information screen in the Australian Technology Park, Sydney.
Source: Author's photo, April 2013.

special infrastructural and technological considerations compared to the other free zones in the city (although some of the larger media companies located there have since situated their own capabilities, but crucially it is owned by the company, not the landlord, as is the case in other media cities around the world).

So cities are constructing these high-specification Media Cities in order to attract the creative economy firms and workers that are so integral to the Creative City paradigm. Seemingly without the office space, the technological capacity, it is unlikely that they will want to locate there. So in order to attract the creative-class workers to live as well as work in these Media Cities, often there will be high-class living accommodation built alongside the creative and media industry facilities. In most cases, these are high-rise condominium style constructions, often located on or by a waterfront (natural or artificial). DR Byen, for example, has a purpose-built large water feature with multi-storey living quarters situated on either side (see Figure 4.8), although unlike other Media Cities the housing is a mixture of private condominiums, student accommodation (mainly catering for international students) and social housing (although how affordable this social housing is to low-income workers is questionable and the constant adjustments to planning regulations make social housing a problematic issue in new developments). Salford's MediaCityUK was purposely located in the Salford Quays, which is the wider plot of land also owned by the landlord, Peel Holdings. Waterfront regeneration has been an intrinsic part of urban regeneration policies for sometime due to the emotional and aesthetic qualities water can bring to

Figure 4.8 DR Byen's artificial water feature with surrounding luxury living quarters.
Source: Author's photo, 26 March, 2012.

urban dwelling (Norcliffe *et al.* 1996), and as such the marketing of Salford Quays as a site for MediaCityUK includes its waterfront location. Hence, high-rise, waterfront-facing apartments surround the main MediaCityUK buildings in Salford, with luxury fittings and services on offer. Most buildings will offer concierge services, high-speed broadband, fully fitted luxury apartments, and other services designed to be as attractive as possible to elite creative-class workers. The proximity to their working environment (no more than a short walk or cycle away) is also viewed as an important characteristic for Media Cities as it limits the experience of the wider city. The provision of luxury accommodation nearby is a deliberate attempt by the planners of these media city sites to create a 'lived' atmosphere, an area that is not solely for work but a place where people can see out their leisure time. Also in Dubai, adjacent to the free zone are high-rise apartment blocks that overlook a man-made water feature, while also only being a short walk away from the beach (although that walk requires going outside of the free zone). These developments are of course not unique to Media Cities; many business parks, large housing developments, out-of-town retails parks and sports stadia have utilised landscaping techniques to beautify the area as a fundamental part of the urban development process. Media Cities are therefore replicating the standardised urban form that represents the contemporary, real estate-led commercialised city, but are doing so in relation to and through a narrative of the Creative City paradigm. This further exemplifies the cyclical and reiterative process of neoliberal urbanisation that the creativity paradigm represents (Peck 2005).

Moreover, the creative industries are predicated upon, and catalysed by, informal networks and social interactions, those that mostly take place in the time after work, usually in the surrounding bars, cafes and nightclubs. As such, not only do the planners of these media cities offer luxury living quarters, but also there are hotels, bars, shops and other auxiliary leisure facilities – all the critical ingredients of a 'successful' creative cluster are spoon-fed. In Salford's MediaCityUK, the planners have offered building space that is occupied by a high-end Holiday Inn hotel, and the space for a number of cafés, food outlets and bars (in addition to the shopping outlet mall located 100 metres away). The provision for this diversity of real estate is then crucial for the planners of Media Cities as it offers the workers the opportunity to interact socially as well as in the work environment, which is of course one of the perceived prerequisites of the Creative City (Landry 2000). The development though is often at odds with the local histories and, with the case of MediaCityUK, do not attempt to engage with reflexive urbanity at all, instead whitewashing the environment with the 'new' style of Creative City development and all the architectural form that it entails:

> Salford's modernised dock areas … are patterned around a generically plush arts centre, studded with glass-walled apartments, opaque offices, marooned shops, chain hotels, the jaggedly liquid Imperial War Museum North, and modernist wire and metal bridges … Sophisticated transformation of industry history was not part of the methodology of those seeking to turn the

original docks into a decorative quayside living, inspired not by dignified working-class Salford, or a sensitively recalibrated post-industrial world, but by Bilbao and Canary Wharf, everywhere and nowhere. (Morley 2013: n.p.)

The eradication of the area's industrial history to create a 'Bilbao and Canary Wharf' is a symptom of the 'cut-and-paste' template system that is so commonplace among this type of Creative City building process. The complete lack of reflexivity of the Media City model means there is little room or desire to incorporate (any) existing community's culture, creative practices and historical nuances beyond a mere pastiche.

As has been noted, these high specification technological capacities and luxury living standards have a substantial cost associated with them, often beyond the reach of local governments. As such, many of these Media Cities around the world are bankrolled and owned by wealthy real estate companies which have been invited by the local authority and governance structures to construct the facilities. As mentioned, Salford's MediaCityUK is owned by Peel Holdings, Dubai Media City is owned by Tecom, an arm of the Dubai government, and DR Byen was developed by Ørestadsselskabet, a corporate entity part-funded by the Copenhagen municipality (although finance did come from DR which brought with it a large round of job losses at DR when budgets overran). However, in some cases, this results in these areas being highly privatised, with strict enforcement of exclusion and spatial laws designed to protect their high investment levels.

During the research visits and visual documentation of some of these Media Cities, there have been instances where private security forces have enforced a no photography law, stating that the land was private, and enforced their own laws independent of the city. In Dubai, for example, I was approached while photographing the architecturally striking ITP Publishing building by a Telecom private security officer, insisting that I stopped taking photographs. (Figure 4.9 is the closest I got before being approached.) Given that this was technically land owned by the Dubai City government, I attempted to inquire as to why taking photographs was forbidden since technically it was public land (despite knowledge of the highly corporatised nature of the Dubai government and their entrepreneurial approach to land management). Despite the security official's excellent English, no explanation was given, merely a repetition of the request to stop. The wider characteristic of Dubai's attempts at attaining 'Global City' status through a highly commercialised and ruthlessly entrepreneurial Foucauldian style of governmentality is well-versed elsewhere (Davis 2006; Kanna 2011) and too rich an argument to go into detail here. Suffice it to say that the privatisation of land in Dubai is near ubiquitous, and access, usage and passage regulations are all strictly enforced.

Private security firms policing urban space are increasingly common across the urban world. However, as Media Cities represent the utopian idyll of the Creative City paradigm – the perfect combination of creative class amenities, real estate development potential and globally marketable urbanity – they are heavily

Figure 4.9 The ITP Publishing Company building at Dubai Media City.
Source: Author's photo.

protected assets. Another incident encountered during the research in MediaCityUK that exemplified the increased securitisation that accompanies privatisation of land was when a mobile food and beverage seller was forced to move a temporary stall from one side of the road (which was owned by Peel) to the other (council land). Also, filming and taking photographs on the main plaza is strictly controlled, with 24-hour notice required if you want to use a film camera, even if it is for educational and non-commercial purposes. Private security personnel often patrol the outside areas and the restriction of public activities (i.e. busking, begging, street vending, protesting, etc.) is much higher than in other local council-operated land. The everyday public use goes on seemingly oblivious to the securitisation, but the constant patrol of security guards in high-visibility jackets emblazoned with the area's logo and the small visual cues (such as the one in the Australian Technology Park in Sydney shown in Figure 4.10) is a continual reminder that these areas are anything but public. The ubiquity of CCTV cameras further crystallises the sense of perpetual panoptic surveillance of these Media Cities, a characteristic that feeds into debates about the increasing 'Fortress City' played out elsewhere (see Davis (1990) for the seminal and prophetic text on the subject).

   To continue with the example of MediaCityUK in Salford, it is clear that the agenda to build this area was one born out of a governmental policy of

Figure 4.10 'Under Video Surveillance' sign in the Australian Technology Park.
Source: Author's photo, April 2013.

capitalising on the popularity of using creativity to catalyse urban development. The political pressure on the BBC to decentralise from London (from both the Labour and incumbent Coalition governments) meant that cities outside the capital began a competitive process to secure what would be a lucrative and highly enviable capture of the UK's national media company and all the rhetoric and meta-narrative of creativity that it entails. Cities such as Bristol, Glasgow, Manchester and Salford put together enticing proposals, but it was the latter two that eventually fought it out for the right to host nearly half of the BBC's production and creative industry capacity (Christophers 2008). Salford's bid, however, included the financial and political weight of Peel Holdings, who owned the entirety of Salford Quays, part of which was the site proposed for MediaCityUK. Salford Quays sits within a wider urban area that includes some of the most deprived wards in the UK (in 2007, Ordsall, the administrative ward of MediaCityUK, was within the 3 per cent of most deprived wards in the UK), and has a predominantly working-class industrial history. As such, Salford's bid was seen as not only financially viable, but it also fitted the rhetoric of 'regeneration' and the potential to catalyse economic growth was an important factor. The narrative from the BBC, Salford City Council and the other many interlinked institutions that collaborated in the development of MediaCityUK was one of creating a Salford that was a 'beautiful, vibrant and prosperous city where people will want to live, work and study for generations to come' (Central Salford, quoted in Christophers 2008: 2318). The language of a Floridian style of creativity was utilised throughout the promotional campaign, and those involved in the bid to bring the BBC to Salford and to create the wider vision of a Media City were very much influenced by Florida's thesis. The pervasive ideology of the creative class and its perceived ability to regenerate ailing local and regional economies then was very much at the forefront of the campaign, exemplifying how the physical manifestation of MediaCityUK is part and parcel of the Creative City progression. By following such a prescriptive urbanity, however, urban governance is enacting policies that actualise the inherent class division of Florida's ideologies. The relentless pursuit of the Creative City utopian ideal has brought about vocal critiques of MediaCityUK, citing a lack of employment opportunities for local Salfordians, a widening social and economic inequality and even a redirection of social services to pay for the maintenance and upkeep. The development of MediaCityUK was highly predicated on the BBC's relocation of some of its high-profile departments and with the promise of the creation of over 2,300 jobs (BBC 2011) for the region. By 2012, of the 680 new jobs that have been created thus far, only 16 have gone to local residents of Salford (Hollingshead 2012). Other critiques include local community groups and representatives of Ordsall who claim that the money spent by the local council to develop MediaCityUK has been diverted from other local services such as libraries and schools (ibid.). Also, as another example, Salford council spent £330,000 on a bus service between Salford Crescent bus station and MediaCityUK. Such an endeavour is highly questionable given its relatively large expense and the concurrent lack of funds the council has for more pressing social services, given it has seen large budget cuts through the national government's austerity programme. Another example includes the demolition of 'Graffiti Palace', a stretch of wall along the Ordsall canal that has been a peri-legal graffiti site on which street artists have been honing their skills. Now though it has been replaced with commercial developments linked to MediaCityUK. Also, the Secret Gardens Festival in 2012 was an attempt to directly engage the local residents with MediaCityUK via creative and digital technologies, yet other than the celebratory event itself, there has been no sustained collaborations (Haywood and McArdle 2012). MediaCityUK also claimed the Carbuncle Cup in 2011, the award given to Britain's ugliest new building.

Media Cities then have quickly become the go-to policy model for cities looking to crystallise their commitment to the Creative City paradigm through large-scale infrastructural developments. Their sheer scale and expense, however, as we have seen have profound effects upon the physicality and sociality of the city. Following the creativity script in this way and the large-scale developments required to follow it through is suggestive of an urban growth model that prioritises property construction and wealth generation via the exposition of the rent gap (Smith 1987). As such, the instrumental mechanisms of the Creative City paradigm are inherently about property development, albeit with the narrative that such development will generate the all important creative 'buzz' that is needed to catalyse economic activity (or so the script reads). But herein lies the contradiction of these creative 'zones' – if they are designed with fostering a creative milieu in mind, by narrowing the provision to an economically determin-istic creativity definition, they are stifling creativity, not stoking it. This is of course dependent on a very different definition of creativity to the one often envisioned by the Florida-inspired Creative City ideologues, one that will become apparent in Part II of the book. However, for now, it is important to understand the effect these creative zones have on the nature of the city more widely, and the critical discourses that exist to negate them.

Summary

The Creative City paradigm demands space, urban space that is designed specifi-cally for creative industry activity and manufactured primarily with the creative class in mind. Leicester, Shanghai, Sheffield, Salford, Dubai, Abu Dhabi, Seoul, Copenhagen and Sydney are the cities that have been touched upon in this chap-ter, but there are many, many others with similar developments. The 'quartering' or 'zoning' of creativity in this way is an instrumental process of the Creative City ideal, in so far as it is mirrors the process of urbanisation. As discussed in this chapter, the private and commercial interests coupled with the urgency of urban councils to stimulate economic activity allows for 'models' of zoned Creative City areas to be created, templates for urban areas that can by copied and pasted all over the world. The homogenising effect of such a process is self-evident, notably in the UK. But more than simple aesthetics, this process of 'cookie-cutter' creative urbanism is problematising the nature of the public–private dialectic within cities. The subjugation of public ownership of urban space for private control is one of the spatial manifestations of the increase in urban entrepreneurialism. Spatial control, as Lefebvre (2009: 188) noted, is a key instrument in the hegemonic politics of urbanisation:

> Space has become for the state a political instrument of primary importance. The state uses space in such a way that it ensures control of places, its strict hierarchy, homogeneity of the whole and the segregation of its parts. It is thus an administratively controlled and even policed space.

So the Creative City paradigm, as well as manipulating the image and exterior perception of a city within a landscape of inter-urban economic competition through aggressive marketing and public relations campaigns (as was discussed in Chapters 2 and 3), it is also wresting control of the city away from the public, into the networks of public-private urban governance. But in this instance, such 'control' refers to the control to define the very nature of what creative practices actually entail. For these creative zones, be they a Media City, Cultural Quarter or piecemeal incubator space, define a certain set of creative practices – those that are conducive to the Creative City paradigm and the economic development agenda that it prescribes. Borén and Young (2013) have argue that creativity is defined differently within and between urban officials, but the critical uptake is always the same, one of political manoeuvring, or 'playing the game', and thereby reproducing across different urban arenas the 'accepted rhetoric of creativity' (ibid.: 1785). There is a sense then that the Creative City paradigm, through the creation of these creative zones, is shaping a creativity discourse that is predicated upon an urban ideology that incorporates a very specific set of characteristics and aesthetics. The shiny glass buildings emblazoned with media screens, the luxury condominiums, the flagship development, the myriad of incubator/creative workspaces (all with brushed zinc and wood finishes of course), the water front cafes and restaurants – there is an aestheticised urban space that is altogether very 'creative'. However, as many of those critical of this narrow, economically deterministic and gentrifying version of the Creative City have argued, it is inherently *uncreative* (Gibson and Kong 2005; Evans 2009; Edensor *et al.* 2010).

As discussed previously in this chapter, the language of such an urbanisation process has been altered in attempts to neutralise the gentrification process and in effect hide the negative impacts that such urbanisation has on the urban areas that are targeted (be in St George's in Leicester, Ordsall in Salford or any other urban area that could generate a large rent gap). Therefore the notion of creativity (and the economic benefits it is believed to bring) is very much part of this linguistic chicanery. The physical manifestation of the Creative City therefore is when the inequalities that it harbours become physically realised. Cultural Quarters, Media Cities, any type of zoning activity that is conducted under the rubric of the Creative City is laden with the social injustices that have been exemplified so far in this book and are part of neoliberalised urbanisation more broadly. In order to attempt to redress such social and urban injustices, it is perhaps pertinent to summarise the arguments of the book so far.

Interlude

The rubble of the Creative City

It is clear by now that the Creative City has been constructed by a suite of intermeshing, often conflicting but fundamentally aligned processes. The shift toward urban entrepreneurialism that Harvey (1989) articulated has been one of the recognisable triggers, but by discussing the impacts of neoliberalism, the Global City, the need for city branding, the political importance of the creative industries, the popularity of the creative class theory and their dovetailing, we have seen how the Creative City paradigm has been realised. The physical manifestation of the Creative City has been exemplified through the detailing of Cultural Quarters and Media Cities, but they represent the more 'identifiable' instances of 'creative urbanism'. The political rhetoric of creativity has a veneer of inclusivity, of an idyllic urban utopia in which simply by unleashing your inner creative self, you will experience cultural vitality, social mobility and, most importantly, economic growth. It is a paradigmatic or even dogmatic creativity. But as has been detailed throughout the book so far, such meta-narratives and political ideology, once it is put into practice, is pregnant with cultural homogenisation and insensitivity, social inequality and immobility, precariousness and wealth generation for the few and not for the majority. Through the aggressive implantation of neoliberal policies of creativity we have seen:

- how Creative City policies have done little to address cultural participation and the recognition of minority cultures;
- how the role of culture in the city is reduced to an economic, tradable good;
- how city branding creates glossy, touristy and unrealistic visions of the city;
- how cultural heritage has become marketised and commercialised;
- how sites of shelter and informal living for homeless people have become expensive apartments;
- how the creative activities of those that don't conform to a creative class definition are marginalised from urban areas;
- how the preciousness of work has increased, and even been glorified;
- how the stringent quantification and subsequent financialisation of subjective cultural practices have come about;
- how identikit creative zones have homogenised cityscapes across the world;
- how there has been a rapid reduction in public urban spaces;

- how the increase in securitisation and privatisation of the urban realm has come about;
- how public funds have been redirected away from services that need them;
- how sterile urban centres have been created that do not reflect the cultural history of the surrounding locality;
- how sites of urban creativity that did not make any profit have been completely removed.

The Creative City is hegemonic in the Gramscian parlance in that it uses soft powers of semiotics, branding and the creation of a system of signs, but also it is hegemonic in a more tangible way through direct and 'hard' means of securitisation, policing and legal frameworks to force people into acting the way it wants.

The Creative City then, as outlined in the Introduction, has a *problem*. It has a systemic inability to stimulate creativity in individuals because it is precisely the antithesis of that. It has a creativity that is dogmatic in that it proliferates and makes profit from the homogenisation of 'creative' urbanisation, and so any attempts to be innovate and produce new forms of urban practice that do not conform to that profit-making process are resisted and marginalised (at least at first). The Creative City has an inability to produce creative urban spaces that encourages alternative, vernacular and 'other' forms of creative practices. This sentiment is neatly articulated by Borén and Young (2013: 1791) who argue;

> ... it is important ... to explore how new conceptual spaces could be created in which policymakers can think differently, outside of their normal professional constraints, perhaps tapping into their mundane experiences and understandings of creativity, exploring their own creativity and engaging them in new forms of interaction with creative practitioners.

However, while it is very much important to explore and experiment with policy discourse and allow policy professionals the opportunities to 'think outside the box', any attempts from *within* a Creative City policy to somehow create an alternative creative urbanity (a new kind of Creative City) will engender the same systemic urbanisation effects that it is attempting to rebuke. As the many critiques of the creativity discourse have argued (and to which this book has so far spoken), this economic determinism is self-serving; the more the idea spreads around the world, the more instrumental and vacuous the notion of creativity becomes. Any attempts to reconfigure the creativity policy discourse *itself* via political economic interventions will only serve to be categorised as 'innovations' in the Creative City paradigm and feed into the replicative model of Creative City urban development. By its very neoliberal nature, the ideology of the Creative City is in *systemic need* of variations, tweaks and alterations to the urbanisation process in order to maintain its aura of novelty and uniqueness that make it so compelling to urban governments. It is imbued with what Boltanski and Chiapello (2005) detail as the new 'spirit' of capitalism (more of which will be discussed in Part II of this book). Attempts at 'creative political thinking' that reconfigure the parameters of the Creative City policy are merely changing the clothing of the urbanisation process. If we continue to tinker with the ways in which the mechanisms of the Creative City paradigm gentrify urban spaces, then the issues of displacement, rising inequality and a narrowing of cultural provision will continue. What is more appropriate therefore is for a total castigation of an economic view of creativity altogether, in favour of a more culturally and socially sensitive view, one that incorporates active urban citizenship rather than passive creative or cultural consumption.

However, to do this requires a new style of thinking creativity that comes from a new set of ontological and epistemological tools. It requires using existing critiques, ideologies and creative urban politics to formulate a new city, one that does not have injustice and inequality built in. To free ourselves from the problems of the Creative City, we need to engage in *lines of flight* and espouse a more creatively equitable city. From the theoretical rubble of the Creative City, we need to realise a creative city. And this is what Part II sets out to do.

Part II

5   Preparing for flight

So, having overseen the deconstruction of the Creative City (with a capital 'C'), what is to be done? Having utilised theoretical, empirical and rhetorical tools to systematically dismantle the paradigm and expose its inequalities, injustices and inconsistencies, can we partake in piecing together a city that celebrates creativity more broadly defined? Can we revel in reassembling the urban in such a way that champions citizenry above consumerism? Can we enjoy envisioning a creative city (with a small 'c')? And what will it mean to be creative in such a city?

The remainder of this book will attempt to answer this through detailing precisely how, through a critical analysis of those creative activities that attempt to espouse a broader remit of urban citizenship, we can begin to ontologically formulate a creative urban fabric, a fabric that defenestrates those mechanisms of inequality, and stitches together all the characteristics that the Creative City marginalises. In order to achieve this though, there is a need to understand that the city that has been deconstructed in Part I was founded upon a singular understanding that neoliberal views engender. The creative city cannot be experienced through such viewpoints. It is multiple, chaotic, playful, emergent, in flux, collaborative. It defies the status quo. Such a city requires a new vocabulary that can be counter-intuitive, antithetical and even nonsensical, but is necessary to comprehend if we are to move toward a more equitable creative city. If we are to evade capture from the hegemony of the Creative City, then we need to prepare the ground for flight. Many of the concepts, ideas, tropes and visions that I will go on to discuss have been well versed in contemporary urban studies and congruent disciplines. (Indeed there have been some accusations of overuse and misrepresentation of many of these ideas.) Yet, despite their literal fortitude in urban studies as individual concepts, as an assemblage, they can elucidate a creative city that looks very different to the one that has been deconstructed. The first part of this chapter will negotiate the language needed to embrace the complexities that characterises a creative city. Luckily, there are linguistic metaphors that ease such a transition, so the tropes of the palimpsest and heteroglossia will be discussed here. Equipped with a vision of a less ordered and defined city, there is a need to revisit some well-known ideas, namely those from Walter Benjamin, the Situationists, De Certeau and related work. By discussing these ideas not only in relation to each other, but also with a fidelity to their original politics, then we can begin to utilise them more readily in emancipating the creativity of the city.

The multiple voices of the city

I once had to commute on a train from Streatham to Elephant and Castle in South London, and I made a point of sitting on the same side every day as I could see a wall that had some intricate graffiti adorning it as the train lay waiting in Herne Hill station. The graffiti seemed to be growing every day with differing styles and colours. However, one morning, I saw a team of four men in overalls with large rollers painting over the graffiti with brick-coloured paint. Despite their best efforts, the faint presence of the graffiti underneath was still slightly visible once they had finished. Then, the following day, the wall had been graffitied once again, but this time, the artists had utilised the faint outlines of the original graffiti to create different and more elaborate designs. They had also used more of the wall, going beyond the re-painted area. A few days later, the team of painters were back, straining with their rollers to reach the graffiti higher up. This toing and froing went on ad infinitum for the entire time I took that particular train journey, a period of about ten months.

The surfaces of the city are multi-layered – one could argue that it is a palimp-sest of past and present stories, cultures and economies etched onto each other over time. This layering of the city by a long history of planners, architects, activ-ists, commentators and ordinary inhabitants creates a multiplicitous canvas which has been concocted by an anachronistic recipe of play, creativity, activism and art. It is not a static canvas though, and it is not a purely unidirectional relation-ship either – the symbiosis of the city with its inhabitants reconfigures these spaces continually. Hence why the term 'palimpsest' can be considered an appro-priate one. Dillon (2007: 4) defines the palimpsest as 'an involuted phenomenon where otherwise unrelated texts are involved and entangled, intricately interwo-ven, interrupting and inhabiting each other.' The urban topography offers count-less examples of this entanglement that have built up over the centuries, with constant demolition and rebuilding ubiquitous in modern city planning processes. There are many examples of explorations of the palimpsestuous nature of cities, with Thomas (2010: 7) offering a particularly articulate one of Prague:

> Prague can perhaps be more aptly compared to a multi-layered manuscript on which numerous writers have left their trace without completely effacing the presence of their predecessors. Envisioning Prague as a palimpsest allows us to understand the city's historical as well as cultural ability to efface all evidence of those who have tried to monopolize it.

By rejecting a monopolising narrative to a city (such as 'Dickens's London or Dostoevsky's St Petersburg), Prague has inspired many poems, stories and plays, but in doing so has etched many layered experiences, memories and ideas – not a meta-narrative that is afforded by a city being the subject of a novel (as Thomas argues, is the case with London, Paris and other European cities).

Other examples of such a palimpsestuous reading of the city (one which reso-nates with my own experiences on the train bound for Elephant and Castle) can be found by Iveson (2012) who, when looking at Sydney's outdoor advertising, found that bus shelters were the sites of different voices and signs, layered on top each other;

> The bus shelter not 100 metres from my office at the University of Sydney is a good example [...]. Despite the daily appearance of a JCDecaux [a major international outdoor advertising company] maintenance van to clean up the bus shelter, it is regularly covered with notices and advertisements for meetings, parties (of both the political and dance variety), and other events. This game of cat-and-mouse is played out on bus shelters in different parts of the city. On other bus shelters, supposedly 'graffiti-proof' glass surfaces are frequently tagged with 'scratchies' etched into their surface. (Iveson 2012: 166)

Another example can be seen in Figure 5.1 which shows a street door in New York and a traffic-light post in Helsinki. In New York, the door has been regularly adorned with postcards, small posters and graffiti tagging. In Helsinki, over the course of time, the post has been adorned with various posters for particular events. As can be seen, each time a new poster is added, the previous one is not removed but simply covered over. In this case, each layer can be exposed as they have been ripped apart, showing the constant addition of messages over time.

Figure 5.1 Multiple layers of advertisements, New York (on the left) and Helsinki (on the right).

Source: Author's photo, 15 May 2013.

Figure 5.2 The rusted staples pole, Queens Street West, Toronto.
Source: Author's photo, 21 June 2012.

Also, Figure 5.2 shows an electricity pylon in Toronto riddled with rusty staples. This speaks to the ghosts of messages that were once attached to the pole, with only the staples used to affix them now visible, along with the odd ripped corner of poster remnants that once adorned it. Moreover, the build-up of staples over time and their rusting has created its own message, its own art form. Local businesses informed me that people now come just to add more staples as well as advertising posters. The continual layering of messages has itself become a message, creating a complex and dynamic intertwinement of textual forms, from advertisement to art form.

This form of layering of messages, the multiple urban stories, is part of the complexity of urban fabric. It is often the case that the non-commercial voices (the lost cat posters or advertisements for club nights that appear on building site hoardings that Iveson points to in the quote above and are alluded to Figures 5.1 and 5.2) are often seen by the city as a nuisance, an annoyance to be covered up at the earliest possibility. They obscure and muddy the urban facade, defacing the original surface that the planners, architects and building managers painstakingly and expensively created. A building that is built for a specific (often commercial) usage has substantial financial investment attached and there are often intensive procedures in place to protect that investment from urban practices which look to use any clean and visible surface to convey a message. As was detailed in Chapter 4, private security firms, CCTV cameras and marginalisation artefacts (such as razor wire) are utilised to isolate a building or other urban space, to keep its original usage in tact, to keep the area 'clean' from these 'exterior' urban messages. The surveillance techniques that have been employed in modern cities, especially CCTV, have an effect of creating a modern-day Panopticon whereby security officials sit in front of a bank of television monitors, akin to the central tower in Bentham's power-centralising Panopticon diagrams. Foucault (2008 [1975]) invoked a 'panopticism' which used Bentham's ideas to describe a 'perfect exercise of power' (ibid.: 185) and this has been further exacerbated by technology which extends the city's capacity for surveillance (Wood 2002). With the proliferation of this type of security activity, the ability then of building managers, companies and security forces to ostracise any activities that do not comply with the specific way in which they want the building or space to function has increased. This, however, does not eradicate the palimpsestuous nature of the place, it merely illegalises the activity of using that site to convey alternative messages, and perhaps even increases the frequency and tenacity of those looking to write alternative messages. After all, the graffiti artists in Elephant and Castle were reacting and adding to the very attempts to silence them. 'Alternative' messages are more often than not though community messages: local voices articulating community-orientated events, pleas for information about a missing possession, instructions by local residents or inspirational quotes and phrases.

Hence, the jostling of messages is constant. This jostling produces a multiplicity of messages and stories that culminate in a cacophony of urban voices. Multiple timings, identities, classes, tastes, places and opinions are represented in this chorus, producing what some have called a *heteroglossia*. Mikhail Bakhtin first used this phrase in 1935 in his book *Discourse in the Novel*, when referring to the multiple voices that can exist in the pages of a novel. By examining the way in which multiple voices are present, Bakhtin has produced an apt mechanism with which to analyse urban life. Without relating to the city initially, it provides a highly appropriate metaphor for the multiplicity that is inherent to the creative city. Moreover, Bakhtin's conceptualisation of heteroglossia was based on the premise that there is no neutral language in a novel and therefore no dominant force, and the different languages and voices will conflict. The non-neutrality of language negates an authoritarian voice, and creates conflicts within novels and with surrounding languages that are unarticulated in the text. Language is then not 'pure', i.e. it comes to us laden with the cultural, historical, dialectical, emotional and personal intentions from which we heard it. It rejects hierarchy. Like two mirrors facing each other and the infinite reflections they produce, heteroglot language forces us to accept the inter-reflecting aspects, the infinitely multi-levelled, multi-horizoned characteristics of that which is being spoken.

The city as a palimpsest allows for the layering of the city to be articulated and a heteroglot city engenders a reading into multiplicitous and non-hierarchical city voices. Often, it will be the official lines of communication that are most prominent in the city. The shop signs, the traffic signalling, the transport announcer, the pedestrian crossing beeping – they juxtapose together to bombard individuals on the streets with the latest commercial message or behavioural impulse. These voices, however, while often viewed as banal and bland, will dominate alternative voices such as the beggar's plea for change, the illegal busker, the British sign language conversation going on in the cafe, the discarded flyer on the floor. But each voice is as intricate to the city's fabric as the next. The diversity of these voices and their cultural, social and personal histories all have something to say. The voices of the Creative City will be commercial, enticing the urbanite to engage in 'creative' consumerism, to continue to contribute to the financial accumulation that characterises much of the creative exchanges in the city. But by conceptualising the palimpsestuous and heteroglot city, these messages, texts and voices are conflicting with others, both heard and unheard. Interactions of a non-commercial purpose are posited in opposition to those of a commercial purpose. The different voices then of the city play against each other, maintaining their meanings by conflicting with the other. Hence the city is awash with these alternative and vastly differing voices, each vying for the attention of the urbanite.

Lefebvre, in his essay *The Social Text* (1961), articulated these 'voices' of the city through a discussion of signals, symbols and signs. Signals 'tell us nothing' and represent the 'redundancy of the modern social text: banality and clarity, triviality and intelligibility' (2006 [1961]: 88). Signs are the carriers of information. Symbols are hidden and transparent, deep-lying information that can surprise, but also form part of the 'background noise' of information – hidden in the overload of urban signage that is so rampant in the Creative City. The mixture of these three make up the social text:

> A good social text, is readable and informative, surprises but does not over-stretch its subject; it teaches them a lot, and constantly, but without overwhelming them, it is easily understood, without being trivial. The richness of a social text is thus measured by its accessible variability: by the wealth of novelty and possibility it offers individuals. (Lefebvre 2006 [1961]: 89)

This social text, the mixing of signals, signs and symbols, is a fine balancing act, and the contemporary city epitomises the interplay between reading and being read, between the imaginary and the unsparingly real. Everyday consumption then envelops (or should that be 'consumes'?) the production of the social text and the city becomes awash with (post-)modernity, signifiers of a fetishism of materialism. The 'hidden' meanings of symbols, however, remain, but they are 'drowned out' by the signals – banality overcomes meaning, consumption trumps all. The economisation of the city then overrides any other use of it. The city's major constructing forces are almost exclusively characterised by capitalism, and those voices that do not contribute to its capitalist accumulation (such as the graffiti on that wall in Herne Hill in South London) are relegated, overridden, marginalised or subsumed.

To say though that they 'disappear' is erroneous, they simply become more of a symbol than a signal (to use Lefebvrian terminology). Instead, it is our *perceptions* of the city via everyday transactions that are almost exclusively orientated toward capitalist accumulation and/or commercial exchange, with the hierarchical voice being the most visible (or audible) and the rest being posited as 'alternative' or 'the other'. The cacophony and complexity of layers, voices and even the smells and sounds that bombard our senses as we go about the city can overwhelm, disorientate and alienate us. Neurological filtering systems attempt to make the everyday bearable, so it is often the brightest, loudest and smelliest experiences that infiltrate those systems and register to us as urban 'reality' – but it is a reality that is characterised by the dominant and hierarchical voices. But instead of limiting our perception of urban 'reality', the non-dominant voices can be sought after and explored. In registering these alternate voices, albeit in a fleeting moment of time, there is a sense of creative interaction within the city beyond that which is prescribed by the Creative City. And this is where the work of the Situationists can help us. Seeking out those non-dominant voices and exploring the cacophony of voices that exist in the city were fundamental to their philosophies. Hence, the following section of this chapter will go on to explain exactly why the Situationists (and the themes and literatures that are commensurate) can aid in the rejection of those Lefebvrian signals and act as a catalyst to the realisation of the heteroglot voices of a different city. There are many different tropes that will be useful in this regard, but it is perhaps pertinent to first discuss one of the most foundational concepts in urban critique, one that spawned a generation of urban scholars intent on resisting the alienation of the modern city – the flâneur.

The flâneur

In Edgar Allan Poe's (1840) *The Man of the Crowd*, the narrator of the short essay is sitting in a coffee shop in an unnamed crowded thoroughfare in London. As he gazes into the crowd, he describes the myriad of different people that pass by the window – from businessmen, noblemen, tradesmen, merchants, balding upper-class men, gamblers, Jewish pedlars, street beggars, invalids, prostitutes, drunkards, porters, organ-grinders, coal-heavers and ragged artisans. The narrator takes delight in his ability to read people with only a fractional glance; that is until he sees an elderly man with an expression that the narrator has not seen before. With a multitude of emotions, histories and past experiences the face of this old man captivates the narrator, so much so that he gets up from his seat, walks out of the coffee shop and proceeds to follow this elderly man. He is dressed in ragged clothing, and continues to walk through the crowded thorough-fare, seemingly without purpose. The narrator continues to follow the elderly man, who apparently unaware that he is being followed. The old man walks through the crowd then into a brightly lit plaza, which he circles again and again. Once the crowds begin to thin and night begins to fall, the old man rushes through another street into a bazaar. The old man wanders from shop to shop, from market stall to market stall, gazing at the paraphernalia on offer, but never buying a thing. When the elderly man is jostled by a shopkeeper, he flees back through alleyways to the busy thoroughfare where he was first spotted. It is now night time and the old man, looking pale, walks toward a theatre, which is now emptying of its patrons. He begins to follow some of the theatre-goers, but once they thin in numbers, he meanders toward a part of city that contains abject poverty, noise and squalor. The vociferousness of life in this part of the city causes the old man to perk up, he is more jovial and begins to walk more briskly. As day breaks, he then walks away from the area, back toward the original thoroughfare where he was first spotted. The narrator, after following him toing and froing through the crowds becomes tired of the aimlessness of the old man's wanderings. The narrator decides to confront the old man, but the elderly wanderer says nothing, and continues on his seemingly aimless walk. The narrator pontificates that the old man refuses to be alone because he has committed a heinous crime of some sort. He concludes that it would be folly to follow, because he could learn no more of the old man's motives, his person or subjectivities by following him.

Can the old man or indeed the narrator in the short story be considered to be what has become known as a flâneur? Walter Benjamin, in his unfinished Arcades Project (started in 1927 and continued until his death in 1940), is one of the main protagonists associated with the description of the flâneur. He drew on (and in some instances critiqued) the work of Charles Baudelaire, both of whom allude heavily to Poe's seminal work. Benjamin describes the way in which the flâneur is a saunterer, a 'man about town'. He 'stands on the threshold of the metropolis' (Benjamin 1999: 10), but yet it does not contain him. He is an idler who is indifferent to the increasing rapidity of modern life. Immersing himself in the sensory offerings of the city, he has no set goals, paths or specific desires, only a want for the pleasures of place and temporariness, experientiality and instantly forgettable topophilia; he basks in the *sillage* of the city. Therefore the flâneur, for Benjamin, is a poetic representation of the desire to engage relationally with an urban context but at the same time remaining distinct from it; the flâneur is always in possession of his individuality. Benjamin posits this against the *badaud*, the 'gawker'. Indulging in 'rubber-necking', the *badaud* throws himself into the fabric of the city and becomes subsumed by dominating forces of the city, he becomes part of the city and the crowd, while the flâneur remains distant because he *records*. He observes and documents like a detective. While on his walk or upon returning to his 'quiet place', he is writing up his experiences like someone who does not yet know what it is he has to solve but is aware that everything could be of significance (Featherstone 1998). For Benjamin, in response to the then-perceived hyper-modernity and rapidity of life of metropolitan Paris, the flâneur can bifurcate into a detective or a gawker; he can detail, observe, calculate and narrate the city, or be subsumed into its totalising forces, or perhaps meander between the two. Indeed, Benjamin (1997: 69) notes that, for the flâneur, 'the joy of watching is triumphant. It can concentrate on observation; the result is the amateur detective. Or it can stagnate in the gaper. Then the flâneur has turned into a *badaud*.'

The narrator in Poe's short story begins as the detective flâneur (Poe subsequently became heralded as the 'inventor' of detective fiction). He is observing the urban crowd through his coffee shop window. Passing comment and critique upon the different typologies of urbanites (some of whom are undoubtedly *badauds*), he is objective, clear and 'triumphantly watching' (as Benjamin would put it) and articulating the heteroglossia of the city. However, upon seeing the old man, the narrator becomes unsettled. He is unable to record and document him as concisely as the others and becomes more and more irate. He then gets up from his table and begins the long and seemingly random trail through the streets of London, always deep in the throng of busy crowds (be it the busy main street, the plaza, the bazaars or the theatre). The narrator has arguably veered from being a detective to characterising a *badaud*, becoming infatuated with the old man's desire to be 'a man of the crowd'. Such insidiousness and clandestine behaviour from the old man is what causes the suspicion in the narrator. And with this, Benjamin notes that Poe's writing has blurred the lines between a flâneur and the act of being asocial, stating that 'the harder a man is to find, the more suspicious he becomes' (Benjamin 1997: 48). Poe's fantastical essay, seeing as it has been utilised in the foundational texts describing the flâneur, is therefore an important account in the history of urban interactions. Its esoteric imagery, investigative prose and distinctly urban narrative characterises it as an urban subversive text ahead of time. Its description of the flâneur is multi-layered and open to multiple interpretations no doubt, but its utility and agency in historicising the epistemologies of urban subversion needs to be underscored.

Benjamin's Arcades Project focused on Paris in the early twentieth century. The substantial pace and rapidity of change in the interwar period in the city and his awe at such a rate of change is clearly evidenced throughout the work. The ways in which the modern city alienates individuals was clearly concerning the thoughts and writings of Benjamin (and many others considered part of the Frankfurt School, such as Herbert Marcuse and Jürgen Habermas, particularly his work on the public sphere) and the increasing grandeur of cities put them more and more beyond the comprehension of a single individual. Much of Benjamin's urban theoretical exploration was developed from a Marxist tradition, developing the class war argument spatially through the urban terrain. Within the contemporary city, specifically the Creative City paradigm detailed in Part I of this book, such concerns about alienation and submission to an overarching and forceful urban governance system are equally as valid, although more ingrained into our experiences of the city. The familiarity (and even perhaps to some comfort) with the overbearing, alienating and subsuming power of the contemporary city, renders its inherent inequalities all the more difficult to expose. However, what the flâneur and associated ideologies provide is a useful starting point with which to pursue a more socially equitable, culturally diverse creative city (with a small 'c').

Unitary urbanism

If Poe's 'man in the crowd' is my starting point of the reaction to the alienation of the city, the Situationists International (SI) articulated its philosophical development.

The SI were established in 1957. They ran a journal until 1969, created and curated artistic endeavours, and engaged in political and urban protests. Consisting of a disparate and varying group of people over the course of their activity, the SI were actively seeking to critique the capitalist system that was building up in the post-Second World War Western world. Their actions, writings and artistic outputs were the result of a realisation of the pervasive encroachment of capitalism upon society as a whole. Based predominantly in Northern Europe (encompassing French, Belgians, Scandinavians and Dutch nationalities), they played a key role, theoretically, philosophically and individually in the May 1968 Parisian uprisings. Their most influential text is arguably Debord's *The Society of the Spectacle* (1983 [1967]), where a comprehensive deconstruction of (then) modern life is undertaken, paying particular attention to the role of media and the image within contemporary capitalistic and urban societies. Ontologically linked to Lefebvre's (2006 [1961]) ideas of the social text discussed previously, Debord argued that the 'spectacle' is the way in which the production and consumption of society has collapsed in on itself, creating a mode of existence whereby the realities of social life are no longer separate from that which is produced and consumed. Through advertising, media, news, entertainment and 'spectacular-ised' consumption (the 'signals' and 'signs' of the city (Lefebvre (2006 [1961])), society can never be separated from the dominant modes of production because it is a part of it. In other words, the image has become the commodity, and they mediate social relations to such an extent that they become the social relations themselves. Debord articulates this most vividly in his sixth paragraph of the book and so is worth quoting in its entirety:

> Understood in its totality, the spectacle is both the outcome and the goal of the dominant mode of production. It is not something added to the real world – not a decorative element, so to speak. On the contrary, it is the very heart of society's real unreality. In all its specific manifestations – news or propaganda, advertising or the actual consumption of entertainment – the spectacle epitomises the prevailing model of social life. It is the omnipresent celebration of a choice already made in the sphere of production, and the consummate result of that choice. In form as in content the spectacle serves as total justification for the conditions and aims of the existing system. It further ensures the permanent presence of that justification, for it governs almost all time spent outside the production process itself. (Debord, 1983 [1967]: thesis 6)

The SI, through their writings, activism, poetry, artistic ideas and philosophy, were therefore attempting to subvert the pervasiveness of the spectacle. From an urban perspective, the spectacle created cities that were instrumental, dominating and alienating. Critiquing earlier modernist urbanism (a la Le Corbusier) as attempting to attain an utopian city that was not yet realised, the vision of the urban that the SI championed was to be born from the potentialities that already existed within the contemporary city. Far from whitewashing the city and starting anew, which was the purview of many urban planners and managers of the time (and still is, albeit in different guises, one of which is the Creative City), the SI were concerned with the possibilities that lay within the pamlimpsest of the city. They championed emancipatory actions that were possible through an exploration of the true self. Many of the artistic, activist and literal endeavours spoke to such actions and fell under the umbrella term of *unitary urbanism*. The SI argued that unitary urbanism 'involved the continuous, conscious and collective recreation of the environment' and was the 'fruit of a new type of creativity' (Pinder 2005: 165). A city that cannot be captured by the spectacle is within reach, but only through a radical change in the behaviour, politics and relationships of urbanites based on the ideas of unitary urbanism. There are many ways in which the SI expressed this critique and articulated such a change in behaviour. The 'mechanisms' of unitary urbanism were diverse, complex and multifaceted (see Pinder (2005) for a far more engaged and sustained analysis), but I want to pick on two that are of particular relevance to the themes expressed in this book: the idioms of Constant's New Babylon with the more general ambiguities of 'play', and the nebulous idea of *dérive* which is linked to psychogeography and the subversiveness of walking.

New Babylon

Constant Nieuwenhuys was a member of the SI and his 'New Babylon' concept was a utopian urban vision based on the total creativity of *homo ludens*, or the 'playful man'. It was the output of a range of activities (writings, lectures, sketches, sculptures, artistic interventions, dérives and so on) that took place between 1956 and 1974. New Babylon was a city theorised to be made entirely by its users; it has architecture of experimentation, or user-generated products rather than being predesigned by architects. It was a city that created situations, experiences that were not prescribed. It envisioned a society that was completely 'free' from the need to produce (as industrial production was completely automated) and therefore free from oppression, inequalities and hegemonic forces (as the bourgeois remained encased underground). New Babylon allows for creativity to be total, as 'the history of humanity has no precedent to offer as an example, because the masses have never been free, that is, freely creative' (Nieuwenhuys 1974: n.p.). The argument put forward is that historically, society has always demanded creativity to progress; the very nature of survival and growth requires creativity. But this puts utilitarian functions to creativity; it assumes that the very act of creation is done for the purposes of development, growth and urbanisation. However, in New Babylon where people are free from the need to create existence or society, creativity is therefore total – we become playful beings (Wigley 1998). In a city where human creativity is not used as perishable fuel for the production of commodities, the act of creativity is a social act, not an economic one. Indeed as Nieuwenhuys (1974: n.p.) notes: 'Among the New Babylonians ... the creative act is also a social act: as a direct intervention in the social world, it elicits an immediate response.' Therefore, in New Babylon, creativity is social, something which creates a public and 'loses its individual character' (ibid.), eschewing the individualistic nature of capitalistic urban development.

In the contemporary city, creative acts that are purely social, collaborative and anti-individualistic are tempered by the forces of the neoliberal creativity rhetoric to 'succeed'. To do so requires individualism, to claim the creative act as part of a 'portfolio'. Therefore acts of pure social and collaborative creativity are crowded out by those where individuals are championed. Indeed, from a methodological perspective, finding them requires becoming a flâneur, to explore the city without the way-finding response elicited by individual encounters. Tompkins Square Park in the Lower East Side of New York City has a history of subversive and anti-authoritarian activity. In 1874, it was the site of political uprisings, but more recently in 1988, there were anti-gentrification riots after the homeless population were forcefully evicted. The area, much like many others in Global Cities across the world, is witnessing massive rent rises, which fuels the need for land to be developed. Many community gardens occupy 'derelict' land within the area, and are used by the local community as allotments, safe havens and places of creativity (Cresswell 2013). One of the largest of such gardens (on East 9th Street and Avenue C) acts as a social hub of creative expression. The fences are adorned with trash-sculptures, where people have artistically used the waste products from the garden and the street (see Figure 5.3).

Figure 5.3 Community garden near Tompkins Square Park.
Source: Author's photo, 24 March 2014.

While officially managed by the city council, the local residents maintain the gardens collaboratively. The aesthetics of the garden (and many more in and around the area) espouse a social creativity that belies any utilitarianism. The 'art from waste' elicits a creativity that is social, anti-individualistic and playful. The efforts of those involved in the garden's creation and maintenance are pockets of expressive creativity in a city of functionalism. Of course, many of these gardens are intersected by practices that seek to align them with a more instrumental and urbanising form of creativity. Indeed, as can be seen from Figure 5.4, the very name, 'The Creative Little Garden', gives the expressive creativity of this location a place in the space of the Creative City discourse. The city council, by merely having its logo on the fence and labelling the garden as 'creative', is arguably shifting the concept of artistic expression within these community gardens away from the 'New Babylonian' style of creativity to be more in tune with the spectacle of the Creative City.

So Nieuwenhuys's concept, while utopian, can give us an understanding of how, being more like *homo ludens*, we can begin to be ideologically free from the co-opting forces of the Creative City. The philosophical yearning for such a utopian idyll and the striving to become the 'playful man' is rife with difficulties and resistance, the most overbearing of which is that being 'playful' in today's parlance is to either adhere to a set of rules (the playing of games) or to revert to

Figure 5.4 The Creative Little Garden, New York.
Source: Author's photo, 24 March 2014.

child-like behaviour. 'Playing' in the Creative City means very specific things. One can be 'playful' in manicured urban public spaces (green spaces, parks, etc.) but these are within certain designated boundaries. Or, one can be 'playful' if it is for a profitable goal. These ideas are developed further in the following chapter, but suffice it to say, 'play' within the Creative City has itself taken on totalitarian characteristics (Langford 2006). However, being playful in the ordered urban totality 'outside' of these prescribed boundaries is an attempt to reanimate the everyday, to create disquiet, ambiguity and insecurity, to be *homo ludens*. Play diversifies urban experience rather than homogenises it because it unsettles the normalising practices of the hegemonic urban forces of the Creative City. Playfulness therefore can be considered the 'starting point' of unitary urbanism and subversion more broadly in that it is ontologically antithetical to the ordered and dogmatic creativity of the Creative City. Hence the importance of grasping the work of the SI and in particular New Babylon, not as a utopian idyll to strive for as an end game, but because it creates the *desire* to challenge, subvert and resist the mechanisms of the Creative City.

In February 1997, a performance artist, Francis Alÿs pushed a block of ice around Mexico City in a video piece called 'the Paradox of Praxis I' (it can be found easily online). The ice block starts off about two feet high, five feet long and six inches thick. Alÿs begins by pushing it, often recoiling at the harm done to his hands. It is hard physical labour to begin with, sometimes lifting it down some stairs. As the block melts it leaves a trail of water along the pavement, and before long, Alÿs is kicking the block along with his feet. After some time, the block is nothing more than an ice cube and he leaves it to melt away on the streets. The tag line for the piece is 'sometimes making something leads to nothing', and is a clear critique of capitalist production. By exerting huge amounts of energy to turn something into nothing is the reverse of the capitalistic ethos. Francis Alÿs' work could be considered to be 'wasteful' in that there is no end product of his labour, nothing to consume. What is perhaps more striking is the urban context of the video. The ice is not always the main focal point of the shot, and the passers by do not seem to notice Alÿs as he toils with the block. This artistic performance does not physically reconfigure the city. It is not an act of reappropriation as there is no 'subversion' or altering of functional usage as such (although perhaps the ice could have been used for commercial or domestic purposes). But it highlights those creative actions which do not yield progress, growth or development. It challenges us to rethink the way in which creative urban activity is viewed as always productive and forever part of the urbanisation process. In New Babylon, creativity is stripped of its productivity and 'agenda-setting'. Similar to Francis Alÿs' performance, it has no progressive political, economic, social or cultural goal or message (although even Alÿs' act has a message, albeit one of critique); it is a more affective and emotional expression of the freedom to act in the city. Hence, the work of SI is critical in implementing and envisioning a creative city as they have given us the conceptual and perhaps even psychological tools to view creativity beyond the economic and urban development frameworks to which it has been aligned so readily by the Creative City discourse.

The subversiveness of walking

Furthermore, New Babylon was originally called Dériville or 'Drift City' and therefore has direct etymological and conceptual links with dérive. At its most basic, dérive (which, as will be become apparent, has conceptual similarity to the flâneur) is moving without a goal, and was formulated most vehemently by Guy Debord. He argued in a dérive, people 'drop' their existing attachments to work, leisure, friends, family and everything that usually compels them to navigate through a city. Then they let themselves be attracted to the urban terrain and immerse themselves in the encounters experienced therein.

Sometimes anglicised to 'drift' (hence 'Drift City') dérive is losing oneself in the city. Extricating yourself from the socialisation of the urban experience, from the constant bombardment of images, signs and messages (i.e. the spectacle), allows for a more visceral form of 'attraction of the terrain', experiencing the urban environment free from meta-narratives and voices other than your own. But more than simply following your subconscious and relying on chance (which was more the purview of the surrealist movement, not the Situationists), dérive is more about being led by the city, an interaction between the human and the inhuman city creating new experiences and new directions to the urban which have not been designated as part of the spectacle. Therefore dérive is about searching for a city *beyond* that which is prescribed and has its functionality, representation and meaning already allocated; it is about encountering a city that is new and unique to that mo(ve)ment, to that situation (Pinder 2005). Dérive is therefore very much a conscious activity to recalibrate and reappropriate the city.

It has its roots within psychogeography, something which Debord himself wrote about in 1955. Specifically though, psychogeography is the 'point at which psychology and geography collide' (Coverley 2012: 1–2) and is more attuned to the affective conditioning of the city, the behavioural changes to the individual that the city can achieve. Enjoying a revival of late particularly in and around London (through writers such as Will Self, Iain Sinclair and Robert MacFarlane), psychogeography is a way of moving and traversing the city (attempting to be) free from cognitive and socialised controls. The act of simply walking around a city (often aimlessly) can be considered subversive in the contemporary transit-orientated city, simply because it is a rare occurrence in modern-day cities where everyone is trying to get somewhere, usually in a hurry. De Certeau (1984), in his celebrated essay 'Walking in the City' argues that walking is when one truly experiences the city. De Certeau obtained a 'God's Eye View' of New York that he obtained from the 110th floor of the World Trade Center (a view not massively dissimilar to that which Phillipe Petit must have seen on 7 August 1974). He argues that to view the city this way to is to be 'lifted out of the city's grasp' (De Certeau 1984: 157). To gaze upon the city is to disentangle yourself from the myriad of experiences, moments and situations that constitute the urban quotidian. De Certeau mirrors many of the arguments put forward by Debord and the SI, arguing that the voyeurism that a gaze from a skyscraper affords creates a hierarchy within our urban experiences. The city becomes an image, a view to be consumed rather than experienced. This view is one that is enacted by, and fuels the creation of maps and plans, a readable city that is coded, ordered and instrumentalist. The city becomes a represented space, which as Lefebvre (1991: 38) noted is 'the space of scientists, planners, urbanists, technocratic subdividers and social engineers'. Such a space lacks any social specificity and practice, but De Certeau goes on to argue that by walking *in* the city, such a represented city dissipates and a more '*migrational* or metaphorical city thus slips into the clear text of the planned and readable city' (ibid.: 158, original emphasis). Walking (or perambulation) therefore enacts a city that defies logical representation, a lived space (Lefebvre, 1991) and therefore can be viewed in and of itself as an act of subversion to the city that is represented by the planners and urban technocrats. Like Poe's man in the crowd and Benjamin's flâneur, walking disrupts normalising tendencies and can react against practices of hegemonic urbanisation.

In March 2013, in the very streets that De Certeau was gazing upon, I met up with Matt Green, who was in the process of embarking on walking every street in the five boroughs of New York City, a route roughly of 8,000 miles. Green has been quoted as saying that 'people tend to narrativize neighborhoods [*sic*] in New York, saying such and such a place is hip, or poor, or ugly, or barren … This walk is a way of understanding a place on its own terms, instead of taking someone else's word for it' (Green, quoted in Lipinski 2012: n.p.). Green's walks have been supported by taking on odd jobs and sleeping in friends' houses or sometimes on the streets. He has been tracking and blogging about the journey and taking the occasional photo to document interesting vignettes of the urban terrain. Green's endeavour can be argued to be part dérive, part flâneur and part psychogeography, as he experienced the entirety of New York's streetscape over the course of the previous three years and counting. By ostracising himself from the 'narratives' that he argues are often pinned to particular parts of New York (cool, unsafe, corporate, peaceful, public and so on), the very act of walking has created his own situations of urbanity that are distinct from those that have been predetermined. From a situationist perspective then, he is experiencing the city on his own terms, creating experiences, ludic encounters and situations that are not conceptualised into an ideology of 'Manhattan', 'New York', 'healthy living', 'exploration' or any other prescribed meta-narrative with an achievable and recognisable concept. Yes, his walk is recorded via the blog, photos and online maps and therefore has been *transcribed*, but however thick or thin, heavy or faint these lines are, they are referring to what has been left after the experience with the city and that street. In other words, it records the *absence* of what has been walked and passed by and cannot capture the experience and situation itself. Via these documentary devices (i.e. being a flâneur that records and documents), Green's walking has been transformed into points, lines and maps similar to those that prescribe the city in the first instance. But after the experience and encounter of the street, it is only relics that are recordable. As De Certeau (1984: 161) has noted, 'itself visible, it [the act of mapping a walk] has the effect of making invisible the operation that made it possible'. Green's walking therefore sits 'in between' the prescribed city of maps and plans (the God's Eye View) and the transcription and documentation of experiences; it is pure experience, pure creation.

Desiring a critique

So far in this chapter, the important conceptual building blocks for a creative city have been advanced. Unitary urbanism and its component ideologies – the flâneur, *homo ludens*, dérive, perambulation – they all offer 'ways in' to emancipating urban citizenry and equipping urbanites with the theoretical impetus to espouse a city of more socially equitable creativity. However, it would be naive to suggest that these are unproblematic, and they should not be used unflinchingly in propositioning creative citizenry. There are a number of critiques that must be explored. By acknowledging their inadequacies, stereotypes and inherent biases, and exposing and testing their theoretical aptitude to radical critiques, we can begin to utilise them more justly and confidently. Once again, there is a plethora of literature to call upon (from a range of political vantage points), and all would make justifiable critiques of the SI, the flâneur, De Certeau and the other associated ideas discussed above. But some are more pertinent than others to the remit of the creative city and are worth developing in further detail.

In relation to the Creative City paradigm that has been described in Part I, it is clear that the arguments of Benjamin, the SI and De Certeau detailed above represent a departure from the neoliberalised version of creativity that has been developed. Unitary urbanism is the practice of envisioning a utopian idyll that champions the innate, social creativity that engenders an urban commons. However, it is important to note that what the SI were critiquing, the society of the spectacle, has shifted dramatically post-May 1968 (Boltanski and Chiapello 2005). The Creative City paradigm, through its structural neoliberal tendencies, has effectively embraced and neutralised such critiques and valorised the creative 'spirit' that can be garnered from unitary urbanism. The ubiquity of the city brand (described in Chapter 2) as part of urban entrepreneurialism encompasses the very critiques that the SI forwarded in the 1950s and 1960s. Innovations in digital technologies of visualisation, way-finding, pervasive media and augmented reality have massively altered the ways in which we consume and (re)produce the spectacle. The 'model' of state power that Debord and the SI were reacting to has radically altered, if not been totally substituted by, a global corporate spectacle. City branding, PR and marketing are the new tools of neoliberal growth, and are designed to seek out difference and celebrate its novelty rather than suppress it. In so doing, it has made activities that engender a diversity of experiences the mechanism of economic hegemony. In other words (as was discussed in Chapter 2) the Creative City actively embraces the celebration of novelty, as it can co-opt, repackage, neutralise and depoliticise.

Moreover, Swyngedouw (2002) argues there has been a sanitisation of the SI in particular, most notably of their urban politics. He argues 'it is as if their [the SI] sting has been removed, as if they have been sapped of the life that once inspired a generation' (ibid.: 154). The confinement of the SI's political agenda to 't-shirts, mugs and serialised postcards' (ibid.) points towards the very logic that the SI were trying to critique. The increasing mediation of society (its politics, economies and cultures) has encompassed the SI and their politics and redacted their revolutionary urban agenda. Unitary urbanism then, it seems, has been conquered by the very spectacle it was trying to subvert. How can we rescue these concepts? Realising a creative city rests upon the utilisation of the concepts of the SI, De Certeau and the like, but if they are readily co-opted by the hegemonic forces of the city, are they of any use? They will simply draw us further into the spectacle, and reproduce the social and cultural problems and inequalities that the Creative City engenders.

To combat such nihilism we can enact Deleuze and Guattari's (1987) notion of desire. Within their magnum opus *Capitalism and Schizophrenia*, Deleuze and Guattari reconfigure the traditional thinking of desire. They move away from a desire that is the want of something that is lacking, a lack that a social configuration entails. Instead they move toward a visceral desire that ruptures those very social configurations. Deleuze's deliberations on desire are woven throughout his philosophy and texts and there is no room to develop the nuances of such work here (see, instead, Goodchild 1996). But if we view the notions of unitary urbanism put forward by the SI through a Deleuzian lens of desire, then there are two things that happen. First, the 'mechanisms' of unitary urbanism – Constant's New Babylon, dérive, psychogeography, walking and other tropes – are stripped of their social configuring politics and viewed as a purer form of a creative desire for a new way of urbanity. Deleuzian desire is a desire beyond that which the conscious self performs. Such 'traditional' notions of desire come about as a direct response to an absence created by a social formation. We want something because the system tells us we don't have it. The systemic injustices and inequalities of the Creative City are 'gaps' in the provision of the urban. Therefore, unitary urbanism can be viewed as a response to those gaps, an attempt to redress the imbalance in an already existing unequal system. However, according to Deleuze, desire is more than simply a response to a lacuna, it is not reactive. It is the creative force for change – it is *desire-production*. As Purcell (2013: 45) has articulated:

> … desire is ontologically primary, it is the source of our creativity, and it produces all things in human society. Accordingly, they [Deleuze and Guattari] use the term 'desiring-production' rather than simply 'desire'.

Desire (or desire-production) is hence the *matter* of all creative expression. An example can be offered by analysing the art of Rowland Matthews, or 'Solider Matt'. A homeless ex-service man living rough on the streets of Camden in London, he would draw ornate, colourful and sometimes harrowing pictures on the pavements around where he slept. Sometimes he would chalk biblical scenes, other times they would be anti-war scenes. He was periodically arrested as his chalking was viewed as anti-social behaviour. Matt died in 2013 at the age of 39. He was an occasional member of a local Anglican church and often talked to many passers-by about his work. One afternoon in June 2012, he stopped me as I walked past to tell me about a scene he had drew which depicted what appeared to be Jesus surrounded by Roman soldiers. I asked him why he had drawn it and he told me it was simply because he wanted to tell people the 'good news of the gospel'. As a homeless man with mental health problems, substance addiction and a criminal record, he 'lacks' plenty within the current capitalistic configuration. Yet he had a desire not to address these things, but to draw quasi-religious scenes on the pavement. His desire produced and created new forms that are far-removed from the system in which he found himself. He was acting creatively not in response to any lack in the system, but out of more visceral desire to produce meaningful new forms.

Deleuze and Guattari (1987) go on to note that there are systems in place to check and 'contain' such desire. Societies, cultures, economies and cities all have 'apparatuses of capture' which suppress desire into reproducing its own modes of existence. Therefore we can see the Creative City as an apparatus of capture that has co-opted creativity and is utilising it to *re*produce itself. Swyngedouw (2002: 154) notes this when he talks of the dilution of the politics of the SI to 't-shirts, mugs and serialised postcards', in that it is the neoliberal creativity paradigm 'capturing' (co-opting) the ideas of unitary urbanism and using them for its own replication. Deleuze and Guattari go on to suggest that we can break away from the apparatus of capture by engaging in *lines of flight*. According to Deleuze and Guattari (1987: 225) 'lines of flight, for their part, never consist in running away from the world, but rather in causing runoffs'. Engaging in lines of flight is to escape capture, but it is more than simply fleeing, as they are always there, they do not come afterwards. They are 'realities; they are very dangerous for societies' (ibid.: 226). Lines of flight are what come about when our desire produces new modes of existence that can damage, subvert or resist dominant ways of being. Indeed, desire-production is always seeking out ways to escape capture. The apparatus counters this by attempting to (re)territorialise those lines of flight as appendages, extensions and 'newness' to co-opt. But by collectivising, being collaborative and social (as New Babylon encourages us), we increase our chances of escaping the apparatus. Therefore by using the concepts of the SI through the lens of a Deleuzian philosophy of desire-production, we can reimagine them free from their apparatus-building qualities that have been layered into them over time. In other words, if we use them to imagine a creative city, then they need to be ontologically removed from their neoliberal tendencies and realigned to their original emancipatory politics. Indeed, as Plant (1992) notes, the SI's agenda has over the years been tainted with negativity and even aligned with capitalist relations. However, she goes on to suggest:

> Even though the ability to control one's own life is lost in the midst of all-pervasive capitalist relations, the demand to do so continues to assert itself, and the situationists were convinced that this demand is encouraged by the increasingly obvious discrepancy between the possibilities awoken by capitalist development and the poverty of their actual use. (Plant 1992: 2)

The 'demand to control one's own life', or desire-production, is a critical component of unitary urbanism. In attempting to realise a creative city then, we should not be put off from utilising the mechanics of unitary urbanism (and associated tropes that I have outlined in this chapter) simply because history has aligned them with the spectacle. We should celebrate community gardens not as part of the Creative City, but as social, playful instances of community creativity. We should explore the experiences of the city on foot and not attempt to label them in an urban meta-narrative. By utilising unitary urbanism techniques with fidelity to their original intentions, then we can begin to start exploring what becoming a creative city might entail.

This does come though with a major health warning, one that requires immediate attention. The flâneur was initially posited as the 'man about town'. Baudelaire's initial concept, stemming from Poe's 'man of the crowd', the flâneur stems from an overtly masculine discourse. The Victorian London of Poe and the interwar Parisian period of Benjamin are vehemently patriarchal societies. If the flâneur is the conceptual ability to extract oneself from the rigidity of urban society, then it is a concept that is only afforded to men (white, middle-class, middle-aged, able-bodied men at that). In detailing the flâneur previously in this chapter, the gawker and the detective are (loose) configurations, but they both inculcate a power to view, a voyeurism that could only be described as masculine. Therefore the flâneur concept is tied to the 'masculine gaze', and as such further developments of the concept also contain such gendered narratives. We see this in contemporary accounts of psychogeography that are mainly narrated by (white, middle-class) men. For example, the headline writers in London offer a highly Situationist-inspired flâneurie account of the city, yet embody the proprietary masculinity of Poe, Baudelaire or Benjamin. The SI did have female members, notably Michèle Bernstein, who was a novelist and for a short period of her life the wife of Guy Debord. Her novels often serialised the sexual freedom that emanated from the politics of the SI, and contained stories and narratives that alluded to the alienating forces of the modern city (Bernstein 2008). But the gendered nature of such alienation is rarely developed further in more recognised SI text and that which analyses it. Moreover, women who 'wander' the city will be scrutinised far more critically than when men partake in dérive, and hence the politics of such acts vary significantly. So far, this chapter has posited that being a flâneur, partaking in dérive and conducting psychogeography encourage us to engage with the city in far less ordered and capitalistic ways. But to do so, to realise a creative city, depends upon us being 'faithful' to the desire-production of their original intentions. However, if we are to do this, then we must realise that their emancipatory qualities are highly gendered. As Part I has detailed, the Creative City and the related creativity paradigms replicate the same hierarchical and gendered tendencies of neoliberal globalised capital in which they are so intertwined (Hubbard 2004). In other words, if the Creative City produces gendered hierarchies to creativity, then there is danger of *re*producing them in the creative city, unless the gender bias of the flâneur and related work is exposed and accounted for.

Similarly, when De Certeau (1984) talks of walking in the city and the situations it can create, it is the experience of walking that is being celebrated. It is escaping the design of the contemporary city which is overseen by technocrats, urbanists and planners, those that that Lefebvre (1991) noted as constructing 'representations of space'. But when such constructions homogenise the physical use of space, it begins to show a bipedal bias (Imrie 2001). Moreover, urban design is concentrated toward those with particular normalised sensorial and mental qualities (Harold 2013). Walking in the city can be liberating and espouse a more affective and performative experience, but the 'migrational' (De Certeau 1984: 158) qualities are only afforded to those whom the city physically allows. Can we realistically utilise De Certeau's work and eulogise its emancipatory experiential politics if the act of 'walking in the city' is painful (emotionally and/or physically)? What if by being mobile, people feel pain? Put even more crudely, can one play if one cannot walk? The answer is of course 'yes', but to think otherwise is an ablist normalisation, which is characteristic of a hegemonic city. But if we return to Deleuze and Guattari's (1987) notion of desire-production, we can utilise the creative forces of all for a more equitable city. For in the same way that the apparatus of capture of the Creative City recodes its critiques as part of its system, it also reproduces its hegemony and its associated gendered, ablist and class biases as 'majority'. Deleuze and Guattari talk of 'becoming-minor', which is a necessary part of desire-production in that to create a new social formation is to move *away* from being the majority. They do not mean this in a numerical sense, but in a more visceral act of uncoupling from the apparatus of capture. By preparing for, and engaging in lines of flight is how we escape capture and begin to realise a new place free from hegemony and related injustices. Becoming-minor then is to essentially reject the majority, that which 'is' and 'has become', and to seek that which is 'becoming'. In 'becoming minor' we eschew any form of majority rule (gender, race, class, body) and seek out that which has yet to formulate into majority rule – that which is minor. In so doing, Deleuze and Guattari are celebrating *de*normalising acts as they are fundamental to desire-production.

Summary

This chapter has prepared the ground for flight by explaining some important conceptual, rhetorical, empirical and ideological tropes which will aid in the realisation of the creative city. By using the metaphors of the palimpsest and heteroglossia, we can envision a city that is multiple, complex, dynamic and multi-layered. And by refracting the politics of the flâneur, unitary urbanism and psychogeography (and other related SI themes) through a Deleuzian lens, we can start to see the layers of the palimpsest, freeing ourselves from the hegemony of the Creative City. However, the question posed at the outset of this chapter, 'what does it mean to be creative in such a city?' remains somewhat unanswered given that the traditional, neoliberal ideas of creativity have been comprehensively deconstructed in Part I. The answer will be become empirically apparent throughout the rest of this book, but theoretically it comes from a further deployment of Deleuzian thought. For if we view the apparatus of capture as the Creative City, then it is by attempting to break away from it, in becoming-minor and engaging in lines of flight, that we are being *truly* creative; we are engaging in desire-production. Conceptually then, it is in reaction to and engaging in lines of flight from the Creative City that we experience *true* creativity because rather than replicate a system of homogeneity, heteronormativity and hegemony, creativity is being used to express difference, form lines of flight, seek new social formations – becoming minor. What does it mean to be creative in the creative city? It means seeking out new ways of *becoming* in the city that are not 'capturable' by the apparatuses of capitalistic accumulation. It means reconfiguring the urban space devoid of the motivation for profit. It means creating a more culturally vibrant and diverse place free from hegemonic political rhetoric. It means enacting more social justice without conquering space. It means constantly probing the 'boundaries' of the Creative City to find alternative ways of experiencing the city.

Creative subcultures, such as parkour, skateboarding, urban exploration, yarn-bombing, street art, flashmobbing, buildering (and so on and so on) are all instances of subcultural activity that have, at some point in their existence, sought to partake in the subversion of the prevailing hegemonic urban discourse. But, like many other activities that have been subsumed by the Creative City, there comes a point at which even these subcultures stop being truly creative and become something else. The next chapter will show us where.

6   Urban subversion

On the evening of 8 April 2012, national media outlets in the UK began running a story about Bradley Garrett, the urban explorer who had posted photos online of his crew's infiltration of the Shard building in London. The photos were from 2010, but by posting them online that day Garrett became the central figure in what was to become a rather fierce media storm over the creative subcultural practice of urban exploration (or urbex as it has been sometimes labelled). All the major UK news outlets ran with the story and he was rushed into TV and radio interviews over the next 48 hours or so. As is the case with these kinds of media outlets, a narrative had to be quickly established, one that could be hawked to its audiences in a bite-sized, pithy and attention-grabbing package. In this case, it was the easy, off-the-shelf, fear-inducing yet spectacularised narrative of terrorism, suggesting that Garrett and the rest of the explorers who scaled the Shard exposed serious security flaws in our high-rise buildings. In the post-9/11 urbicide imaginary, spectacular footage from people who have 'sneaked in' to the then-building site exposing the security flaws make for a highly publishable and broadcast-amenable story. The spectacular photos and videos that Garrett had obtained were splashed all over the papers and television, inspiring awe, fear and outrage all at once. In these highly mediated narratives, Garrett himself was attacked from both 'sides'. He was derided from the urban exploration community as a 'sell out' invoking vulgar invective online from other people who claimed he had made urban authorities more 'aware' of urban exploration and therefore the practising of it more difficult. The mainstream media also pictured him as deviant, a quasi-criminal who was trespassing on private property with no regard for health and safety laws. The reality, however, was of course neither of these, as a quick read and comprehension of his published material (academic papers, blogs and even his PhD thesis) would have shown. Instead, however, the media (both the mainstream and that emanating from other explorers) continued to portray this particular group of urban explorers as thrill-seeking nuisance deviants or money-grabbing sell-outs. There was no room for anything in between.

Within the contemporary Creative City that is governed by a neoliberal governance system, actions like that undertaken by Garrett and other urban explorers are controversial because they defy the prevailing narrative. Why would people risk their lives to climb Europe's tallest building site if not for larger geopolitical or criminal purposes? Why would a group of mainly young, white, middle-class, able-bodied men risk arrest and criminal persecution if not for notoriety and the financial rewards that it can reap? Garrett himself, in many of the television, radio and online interviews, spoke highly eloquently and concisely about how urban exploration is neither of these polar narratives. He often spoke of the desire to experience the city beyond official guidelines, to engage in modern day psychogeography and dérive, to subvert the meaning of place, to write forgotten histories and highlight the crass consumerism of the modern-day city. All of these (and perhaps none of them) are reasons and motivations to conduct the kind of activities through which you can find yourself the centre of a media storm, but from within the context of the contemporary Creative City, there is only need for one – the want to appropriate activities that are as of yet not commercialised. A quick reading of urban exploration as a subcultural practice will highlight its resistance to such commercialisation and its subversion of hegemonic urban control. It shows how it can be considered a line of flight from the apparatus of capture. It can be considered, then, very much urban subversion.

But what is 'urban subversion'? By the end of this chapter it will be clear that the term has both opportunities and hindrances when applied to the kind of activities that can help us realise a creative city. In 2008, I started a Twitter account called *@UrbanSubversion* simply as a curatorial tool for the myriad of examples of people who were engaging with their city creatively. Sparked off by my interest in parkour (more of which is outlined later in this chapter and the next), I came across evidence of more and more people reappropriating the urban terrain in exciting, different and creative ways – young men skating on bank plazas in Madrid when the bank is closed; a student wrapping trees in colourful yarn in Chicago; parkour practitioners (called traceurs) jumping over walls and benches in Vauxhall, South London; performance artists suddenly standing statuesquely still in a busy New York street; communities making street furniture out of old rubbish in Bogota; pranksters modifying street signs to say something political, comedic or affectionate; and many, many more. The reasoning and justification for these people changing their city in ways it was not originally intended was sometimes not immediately evident, yet the very fact that someone took the effort to reconfigure their city to express a particular belief or opinion was enough for it to be considered 'an urban subversion'. In the intervening six years, there has been a noticeable change in the contextualisation of such activities, with a rapid rise in the collectivisation of activities into identifiable (subcultural) groupings. Skateboarding, for example, rose to prominence in the 1960s and 1970s to become an internationally practised activity (Borden 2001), and latterly we have seen parkour become a globally recognised pastime and, with Garrett's actions, we can now talk of 'urbex' as a 'thing'. Often posited as resistive to 'official' rhetoric, they espouse a counter-cultural ethos that has embroiled within it a subversive politics that seeks to challenge the Creative City and its capitalistic operations. Hence, these collectivised, creative subcultural pastimes can also be labelled as 'urban subversions'.

Or can they? With the Creative City paradigm debunked in Part I and a more appropriate suite of conceptual tropes for proceeding toward a more socially just and equitable city identified in the previous chapter, it is important that we proceed with caution. If we are to embark upon lines of flight and be *truly* creative, then we need to be mindful of valorising and indeed fetishising those activities which are in themselves identifiable as somehow resistant to capitalist accumulation. Urban exploration and the burgeoning literature catalysing its transmogrification into an identifiable pastime (Garrett 2013; Gates 2013) is part of a growing trend of recognising subcultural activities that are 'creative' because it engenders a more open, explorative and experiential city – one that was envisioned by Benjamin, Debord, De Certeau and the like. Their emancipatory, subversive and/or transgressive politics have been the subject of notable and high-profile debates of late (in academia, online and in the mainstream media). Much of this debate mirrors the polar mediations of Garrett's infiltration of the Shard. On the one hand, some commentators have eulogised how many of these subcultures are 'markers in the sand' against capitalism. They represent how there are still pockets of resistance in the city that are using artistic practices, virtual media and subversive actions to rally against the injustices and inequalities that are inherent in the Creative City. Other lines of argument have countered this by detailing how these subcultures are just new forms of capitalistic accumulation. They point towards how popular culture (films, television, music videos, advertising campaigns, etc.) have all embraced the 'new' forms of creative expression that are to be found within urban subcultures and that they are themselves branded by multinational corporations. Moreover, cities are creating spaces that are specifically built for such activity (more of which is detailed in the next chapter) and so, fundamentally, there is an accusation that many of these creative subcultures have 'sold out'. However, by utilising and building upon the ideologies from the previous chapter, we can begin to forge a city that celebrates creative subcultures as neither of these two poles, but as communities of practice that are always in flux, always looking for new forms of expression, always seeking to flee from the apparatus of capture, always rehashing and subverting the existing city to create new instances for cultural, social and political expression.

It is also worth highlighting that the term 'urban subversion' was not designed to be a direct pseudonym for urban subcultures. What I have outlined elsewhere (Daskalaki and Mould 2013) is that the term should designate those urban social formations that are temporary, fluid and fleeting – the momentary crystallisation of creative practices that reappropriate the urban topology in innovate and unexpected ways. It is only through the collective and social actions of practitioners and the slow diffusion of the cultural products that they produce (be they stunning photographs of derelict buildings or construction sites, or bright and colourful wool adorning a phone box) into 'mainstream' society that renders the activities more ossified and hence identifiable as a subculture. The continued performance of particular activities breeds certain trends, practices, even rules which coagulate disparate (geographically, socially and politically) participants into a cohesive 'whole' (and, of course, the virtual dissemination of mediated aesthetics aids and short-circuits this process). The modern city is home to a vast number of identifiable urban subcultures, but they are not urban subversions. Often starting out as creative reappropriation of the city (e.g. skateboarders using empty swimming pools in people's back gardens in Los Angeles in the 1960s and 1970s (Borden 2001)), subcultures have proliferated in the age of increased (virtual) connectivity. The ability of people to share in 'communities of practice' (Wenger 1998) creates an identifiable 'subculture'. Often, these people will be practising similar activities completely unaware of each other, yet over time (and aided through online presences) they will be drawn together into more and more cohesive groups. This chapter will hence detail this process and use some of the more 'identifiable' subcultures to exemplify particular points, explain their histories and internal tensions and explore their justifications. They are not meant to be in-depth qualitative accounts – there are plenty of others that have undertaken fascinating, performative, explorative, lived and brilliantly illuminating ethnographic work into specific subcultures. But I have undertaken many of them more than recreationally, and can speak more theoretically to how they can help (or hinder) a creative city. At the end of the last chapter, the question 'what does it mean to be creative?' was answered by delving into Deleuzian-inspired ideology and seeking out those lines of flight and those instances of reaction to the Creative City. The remainder of this chapter then will explore how this can be articulated *epistemologically*. In other words, if a Deleuzian ideology can help us realise a creative city, we need to know how this 'looks' in relation to the city. Theorising a creative city through relevant ideological tropes is only the first step – we need to understand how this theoretical mantra translates physically, emotionally and virtually. Thus the first part of this chapter takes us to the level of the urban object.

The act of creation

The topology of the built environment is an assemblage of individual objects. Walls, benches, pavements, roofs, bollards, bins, gates, stairwells, fences, railings, ramps – the list is endless. These items would have been built, made and placed for a specific reason in the first instance. They are physical manifestations of what Lefebvre (1991: 38) noted as 'the space of scientists, planners, urbanists, technocratic subdividers and social engineers' – the represented space of the city (the space that De Certeau encourages us to defy by simply walking). A stairwell that connects the street level to a subway underpass serves to allow people to descend under the street level and enter the subterranean passage with relative ease (if physically able to do so). Fairly simply then, that stairwell's function is to assist (bipedal) movement and allows urban mobility. However, how would a performative artist view the staircase? Or a traceur? Or a trial-rider? Or a skateboarder? Its 'function' would be very different – one of assisting performance and expression. The stairwell becomes a stage on which to perform a move. The stairwell's function then is dependent upon the *system* in which it is being utilised, which is true of any artefact within the urban topology. More often that not, however, the overarching system that these artefacts align to is one of capitalist appropriation. If we adhere to the metaphysics of objects forwarded by Baudrillard (1996: 67) then 'every object claims to be function'. The function of an object, however, is not the same as its functionality:

> With reference to 'function' it suggests that the object fulfils itself in the precision of its relationship to the real world and to human needs. But ... *'functional' in no way qualifies what is adapted to a goal, merely what is adapted to an order or system:* functionality is the ability to become integrated into an overall scheme. An object's functionality is the very thing that enables it to transcend its main 'function' in the direction of a secondary one, to play a part, to become a combining element, an adjustable item, within a universe of signs. (Baudrillard 1996: 67, original emphasis)

Urban technological objects then, have a *functionality* that transcends their *function* and creates a 'secondary one'. Baudrillard expands upon this by arguing that the functionality of an object makes it part of a 'universe of signs' that constantly mediates between the object's materiality, and the desire to consume it. In other words, an urban object's function (its ability to satisfy our needs and desires) is transcended and obscured by its functionality, which is derived from a system of signs that is constructed, maintained and constantly manipulated by urban governance and Lefebvrian technocrats – it is the space of the city being represented to us. Baudrillard's universe of signs can be articulated as part of the society of the spectacle: 'In form as in content the spectacle serves as total justification for the conditions and aims of the existing system. It further ensures the permanent presence of that justification, for it governs almost all time spent outside the production process itself' (Debord 1967: 6). This is why it is critical to have an understanding of unitary urbanism's critique of the spectacle as it serves to give theoretical justification to attempts to realise the *function* of urban objects (and not be seduced by the functionalities that it is related to). Baudrillard then is suggesting that the 'function' of an object, i.e. its ability to satisfy any particular desire we may have (what he calls the 'order of Nature' (Baudriallard 1996: 68)), is transcended and denied, making 'the system into a system of disavowal, lack and camouflage' (ibid.: 69), or the spectacle. We can also read this as urban hegemonic forces clouding the *function* of an object with *functionality* to a system of capitalism, mediating constantly to us how a particular object should be seen, used or consumed.

However, by not succumbing to the prescribed functionality, by 'seeing through' the camouflage of the urban system, by reading further layers of the urban palimpsest, we can begin to realise the object's function in relation to a particular need. We can therefore again return to Deleuze and Guattari's (1987) notion of desire, as it articulates the means by which urbanites realise the function of a particular object in relation to a *desire* to express a particular emotion, belief, affect or opinion. As I mentioned in the previous chapter, theoretically, desire is the matter of all creative expression, it is creativity that is not 'responding' to a lack in the system, it is not the desire to own a new house, a bigger car, the latest fashion accessory, it is the fundamental desire to create different modes of becoming, to flee from the status quo: it is *desire-production*. What this means from an empirical perspective therefore is the desire to realise the function of an object regardless of any imposed functionality (by technocrats, the Creative City, capitalism and so on). It is the desire to produce a new mode of becoming by engaging with that object regardless of what purpose it was made for. By discovering this function, we can begin to see the act of creativity. It is not necessarily subverting the functionality of an object and therefore reappropriating it for an alternative politics (this often comes later), but merely seeking the object's function as satisfying a particular desire that is not adapted to a system of signs or coming about reactively to a lacuna in that system. An object's functionality, as Baudrillard (1996) argues, comes from its adaptation to a 'universe of signs' that is maintained by an ordered system. Most readily in the built environment, that order is one of urbanisation – be it facilitating mobility, consumption, growth or capitalistic development more broadly (or the Debordian spectacle). We can say that an object within the urban terrain has been placed there for a specific reason, a context that is overwhelmingly one that catalyses urban development. It is in realising the different *functional* possibilities of an object that do not have any context or reason that we can begin to see creativity in its more visceral, affective form.

Let me offer an example. In my (all too brief) ethnographic study of parkour, I often saw traceurs practising and honing their skills in the stairwells and underpasses that surround the IMAX cinema on the South Bank in London. If a traceur was using a stairwell to perform parkour, the stairwell's functionality as a subterranean passageway is not evident. The stairwell's wall becomes an object that facilities her corporeal desire for movement, it is performing functionally in her desire to perform parkour. The stairwell becomes a vaulting device, an obstacle to overcome and even a barrier all at once. Also, perhaps she injures herself while practising the move for a film (something which happened regularly), the staircase then appropriates a function of fear, a pain-inflicting obstacle. If the injury is particularly bad or mobility threatening, then the staircase becomes a site of lasting psychological significance. Conversely, it could also become a site of triumph. All these functions of desire are *possibilities* that are imbued not in the materiality of the object itself, but in the connections and assemblages that people and things make with it and the processes by which they come about. Hence, then, the stairwells and passageways around the IMAX in London are built in the city for a particular functionality, but have an infinite amount of functional possibilities that are unleashed in whatever way people decide connect with it (artistically, corporeally, subversively or otherwise). Some manifestations may be more rare than others; it may regularly become a performance area for parkour, it may become a site for a major artistic intervention once or twice, it is highly unlikely to ever be the site of a major musical festival. But all functions of desire are indeed possible, just some are more possible than others.

Realising some of these alternative possible functions will require more effort, resources and/or time, some only need luck, serendipity and/or chance. But in realising these potentialities physically, virtually or affectively, we can begin to see *true* creativity as it destabilises and subverts the functionality of an object from its designated systemic order (or capitalistic urbanisation) to be more open, varied and non-contextualised. We can begin to see how subversion (of functionality) plays a crucial role in creativity. Being creative in a creative city is searching for these possibilities that are as 'far removed' from the imposed *functionality* as they can be.

Marginalisation

Such creativity though is not part of the Creative City paradigm. Such a dogmatic view of creativity relies on masking the functional desire-production of urban artefacts to a functionality that is restricted to that which the technocrats initially programmed; any realisation of alternative possible functions must be sanctioned to the wider system of capitalistic accumulation and urban development. It relies on the topology of the built environment being utilised in the way it was designed to by urban technocrats; any reuse or reappropriation of that object needs to be validated by hegemonic urban governance. This is because by remaining used in a way that is designated by its functionality to a wider system of order and rules, the object is used *predictably*. Creative activity that realises different possible functions of that object, does not relate to the designated functionality, and therefore cannot be regulated, controlled, predicted or ultimately profited from. Within the inherent neoliberalised and capitalistic logic of the Creative City is an attempt to regulate activity that is not amenable to profiteering and direct or indirect financial reward. Moreover, such regulation comes about through *marginalisation*, the attempt to remove any 'unwanted' or unprofitable activity from the city, in other words to actively restrict the realisation of functional possibility and maintain a specific functionality. This is done very practically in the modern contemporary Creative City. Signs and specific architectural artefacts that prohibit or restrict activities which use a building, public space or urban object in a non-sanctioned way are everywhere in the city. 'No skating or skateboarding' signs can be found in many cities across the globe (see Figure 6.1) given the relatively long history of skateboarding being a nuisance to urban officialdom.

Other specific examples include the process of 'knobbing' where small metal knobs are placed at regular intervals along smooth granite or metal blocks in order to stop skateboarders or urban trail riders using that particular urban artefact (see Figure 6.2).

Also, anti-graffiti paint is often utilised as a common way in which businesses and urban councils attempt to deter tagging and street art. Indeed, there are a number of how-to guides that have been written about the best way to deter graffiti in the urban environment (Whitford 1992). Increasingly, though, we are seeing signage that is prohibiting 'parkour-like activities' and 'urban sports' (Minogue 2014) indicating that the city's marginalisation procedures are adapting to the

Figure 6.1 No skateboarding signs in Toronto and Manchester.
Source: Author's photo, 19 June 2012 and 28 March 2011.

Figure 6.2 Anti-skateboarding 'knobs' on an urban object, Sydney.
Source: Author's photo, 16 April 2013.

Figure 6.3 Signage prohibiting alternative functions at the Sydney Technology Park, Sydney.

Source: Author's photo, 15 April 2013.

'newer' subcultures. The Cultural Quarters and Media Cities detailed in Chapter 4 are often riddled with such signage (it forms part of the wider issue of defensible space (Newman 1996)). Figure 6.3 shows the sign 'please do not climb on crane' adorning the relic of an old warehouse crane in Sydney's Technology Park that symbolises the area's industrial heritage (in a way that echoes Montgomery's (2003, 2008) insistence that heritage be used 'emblematically'). It is there to deter people realising its alternative possible functions as a climbing frame, a site of play, political protest and so on. The addendum to the sign 'This area is under 24 hour surveillance' is also symptomatic of the panoptic nature of many Media Cities around the world, and offers a rather threatening overtone to the seemingly polite request.

So, effectively, those who are managing urban places that are often frequented by skaters, traceurs, graffiti artists and so on are illegalising these activities and reinforcing the Baudrillardian functionality of that building/urban space to the Creative City system. In some cases, it would seem that the prohibition of these activities is based on legitimate concerns about the safety of those individuals who would seek to partake in the activity, and indeed any potential spectators or passers-by. However, the health and safety frameworks that are part and parcel of urban management are also a part of the governance structures of the Creative City. It is but one of the bureaucratic processes of the urban neoliberal assemblage (what could be said to be another method of Foucauldian governmentality). The legitimacy of health and safety concerns over 'improper' use is simply another method of control over the use of a particular urban object. It remains the case that city officialdom displaces these activities, thereby marginalising them. The constant rebuking of alternative *functions* of objects throughout the city begins to create a dualistic logic of *functionality*, with official and formulaic processes on the one hand, and those that do not conform to this on the other, causing an inevitable schism. So, instead of opening up new possibilities of urban function, many activities are set against the Creative City as deviant or 'other'. They have been marginalised.

Acts of creativity that realise other potential functions of urban objects are, however, seldom stopped by such signage and overt marginalisation tactics. They are simply displaced. The 'realisation' goes on elsewhere: physically in other less policed and watched areas and conceptually 'outside' the system of capitalistic urbanisation that the functionality of that object adheres to. As they have been marginalised, they are hence posited in opposition to the Creative City. Yet despite their contextualisation as opposite, those who realise a function of an object that speaks to their desire do not yet 'view' themselves as opposite, they are simply indulging in their desire. Following on from the example of the crane in Sydney, when sat having a coffee in the same plaza, I witnessed a child who climbed onto the crane. There was no indication that the child (who could have been no more than six years old) was climbing on the crane in reaction to the capitalistic devices of Sydney's Media City, she was simply enjoying herself. She was fulfilling her *desire* to use that object as a function of play (perhaps being a citizen of New Babylon). She circumnavigated the functionality of the crane as an object to be viewed, to be admired as a piece of emblematic industrial heritage, and realised its function as a plaything. Much like Solider Matt's desire to draw on the pavement, the child's desire was not directed at addressing a lack in her life, it was not achieving some goal that perpetuated the system to which the crane's functionality is a part (i.e. the Technology Park, the City of Sydney, industrial heritage, cultural landscape and so on). It was Deleuzian desire – a more affective connection between the child and the crane devoid of signs or systems. It was a realisation of an alternative function, one possible use in a universe of uses. This is a rather incidental vignette, but nevertheless it highlights how alternative possible functions, when realised, are expressions of desire. She was playing for no more than about 30 seconds on the base of the crane, and no official from the Technology Park (many of whom I had noticed subsequently on patrol) came to stop her. It was in fact her father who came to drag her off the crane, shouting at her and telling her off as he did so.

Such pure desire could be said to have been a line of flight from the apparatus of capture. The 'flight' was recaptured, it would appear, not by the apparatus of the city, but by the family unit. As was discussed in the previous chapter, Deleuze and Guattari (1987) assert that escaping the apparatus requires combination – put bluntly, strength in numbers. If one person realises an alternative potential function of an urban object and subverts its functionality, they can easily be admonished. But the more people that are involved, the stronger the line of flight's resolve, the more *intense* it becomes. Sociality then (as is the case in New Babylon) is an integral characteristic of subversive creativity in the city. Being social and collaborative can catalyse the realisation of alternative functions. The more people continue to perform the different functions (be it vaulting a wall near the IMAX in London or climbing a crane in Sydney), the more intense that function becomes. It is in this collaborative action that we begin to see the creation of subcultures. The term 'subculture' though is itself problematic, and I prefer to talk more of the process of *subculturalisation* (Daskalaki and Mould 2013).

Subculturalisation

In attempting to analyse the social aspects of urbanism, Claude Fischer (1975: 1320) argued that subcultures come about through the 'congregation of numbers of persons [*sic*]' which is 'sufficient to maintain viable unconventional subcultures' (ibid.). The argument he makes is 'nonecological', which says that 'the source of social action is found within the small milieus of personal life' (ibid.: 1323). He was rebuking Wirth's (1938) notion that the congregation of people in urban centres leads to the weakening of interpersonal ties, instead arguing that cities created a number of smaller subcultural groupings that are collectivised around a certain type of (often deviant) behaviour, which we can articulate now as the realisation of alternative functions. Since then, the term subculture has been used throughout urban sociological literature to denote particular groupings of people around a specific activity, be that specific types of music, sexuality or belonging to a particular gang identity. Maffesolli's (1995) famous work on neotribalism furthered the 'subcultural' model that had been incumbent in sociological literature, presenting a model of urban social activity that sought to explain deviance, counter-cultural activity and, by extension, subversive behaviour across the world. (Incidentally, given the understanding of neotribalism, it could be argued that the creative class thesis bears a striking resemblance.) The use of neotribalism in lieu of subcultures, however, has somewhat negated the experience of place (Bennett 1999), notably the city and hence the very term 'subculture' as Fischer (1975) first articulated it can still hold some validity today. Also, Hebdige (1979) saw subcultures as active forms of resistance to dominant forces, adding a subversive dimension to Fischer's (1975) more objective description. Others have more recently added the prefix 'creative' to form creative subcultures (Shaw 2013), which indicates those subcultural activities that are based on music, art, theatre, old and new media. These are also posited as having an 'alternative politics' and are often at odds within mainstream or dominant cultures of consumption.

Urban subcultures (those that can be identified at least, such as parkour, urbex, yarn-bombing, flashmobbing, etc.) can be said to be another 'incarnation' of the subcultural model, in that they are groupings of people that are collectivised and therefore made manifest by their similar actions within the city. However, they can be seen as creative, not because they engage with the kind of creative activity that is part of the prevailing creativity paradigm (film, art, music and so on), but because they are realising alternative possibilities and potential functions of the built environment. Undoubtedly, 'traditional' creative practices such as film-making, music, art, theatre and new media are part of their proliferation and celebration. But these are post the event of creation; they exacerbate them into socialised, contextualised and tangible formations, but they do not account for the more visceral, affective act of creation that has been detailed in the previous section. They are instead, collectivised and articulable formations of what were once lines of flight.

So, subculturalisation as a process starts with the marginalisation by urban governance and continues through the sociality of those who have been marginalised. Such sociality means that many of these activities will congregate into groups: people who have experienced similar marginalisation will come together (physically and/or virtually). The presence of a community of participants, like other epistemic communities in economic and societal groups inevitably creates a set of practices which are, over time, formalised and adhered to by others (Knorr Cetina 2001; Wenger 1998). Media, Internet and communication technologies further facilitate the formalisation process through dissemination (e.g. tutorial videos and forums). Subsequently, what starts out as a group of friends climbing on walls can turn into a global phenomenon over the course of a few years (as was the case with the emergence of parkour (Mould 2009)). Furthermore, Borden (2001: 137) has articulated the oppositional nature of a skateboarders' subcultural status to conventional codes of behaviour such as the family: 'groups such as the family which is a clear symptom of a set of alternative urban practices becoming more solidified as a formalised urban subculture'. Therefore, belonging to a subculture means that members adhere to a particular identity (for example that of the 'skateboarder', 'traceur', 'urban explorer', 'graffiti artist') and often they will have the need to be recognised by others as well as members of their particular community. And it is here that we see the 'end' point of subculturalisation: the gradual formation of identifiable subcultures through the circulation and dissemination of that particular identity. We see the formation of another system, other sets of rules, formulations and frameworks, that operate as an apparatus of capture. We see the original subversion of urban functionality (and realisation of alternative functions through desire-production) transmogrify into another functionality, hence the subculture as an identifiable 'thing'.

This is a critical point – closure and the cessation of becoming. Once we can 'see' a subculture it invariably succumbs to a dualistic ontology. It ceases to be *truly* creative because it is no longer realising other possible functions to the built environment and subverting functionality, it is instead creating its own systems of functionality. It is creating its own signs and symbols, identities and logics, politics and even its own apparatuses of capture. It has subverted the established order of one system of functionality for another. An identified subculture begins to mirror the structure of the Creative City and risks generating similar inequalities. The politics of that subculture will play a role in determining whether or not it takes on more resistant characteristics (such as skateboarding and urbex) or becomes a subculture that is more vulnerable to consumerist tendencies (such as parkour, flashmobbing and street art). This is a harsh and rather crude delineation as subcultures have internal schisms, debates, tensions and infighting (as we saw with the invective experienced by Garrett), but such delineation is epistemologically necessary as it allows us to further explore an understanding of how a creative city can be realised and also explore the spaces of *becoming* that are 'in-between' these two *beings*. Moreover, it is the politics of subcultures that are in need of unpacking as they can sometimes obfuscate the realisation of a creative city through their own political hegemony and can risk duplicating existing inequalities. It is to these politics of subversion that we know turn.

Subverting the Creative City

One of the reasons for opening this chapter with the account of Garrett's infiltration of the Shard is that it exemplifies how a contemporary subculture, urban exploration, can be misinterpreted by mainstream and online medias. This is because Garrett has written extensively about how 'urban exploration *can be* read as a reactionary practice working to take place back from exclusionary private and government forces, to redemocratise spaces urban inhabitants have lost control over' (Garrett 2013: 4, my emphasis). Indeed, it *can be* a process of subversion that has the deliberate goal of reclaiming private urban spaces for more common public worth. It can be then an act of deliberate resistance. Urban exploration, as a subculture has a vast online community that has its own code of ethics, rules of engagement, dos and don'ts and so on (see Ninjalicious 2005 for one of the first of these). The accounts of infiltrations, break-ins and subversive explorations that have been published so far (see also Gates 2013) all highlight the contested but nevertheless subversive nature of their activity. Garrett (2013) calls the practice 'place hacking' and inculcates a subversive politics that asserts a 'spatial freedom' from the increasing securitisation and restrictive usage of the city. So the prevailing politics is one of reclaiming the space of the city. This carries much of the theoretical inclination of Debord and the SI more broadly, along with many other of the resonating ideas outlined in the previous chapter. Bennett (2011) has suggested that these ideas, which he has called political psychogeography, align with urban exploration and the desire of those who partake in it to recognise a more open and experiential city. But he is more considered in his approach to urban exploration, calling out those who see urban exploration as subversive. When studying a particular subset of urban explorers, which he calls 'bunkerologists' (those who explore disused bunkers), he found something different:

> There is little evidence that ... bunkerology is actually motivated by a desire to perform resistance, transgressive, or alternative readings of these

abandoned places. Instead … the prevailing attitude is more positivistic, reverential, and reconstructive. (Bennett 2011: 432)

There has been some debate between what Bennett sees as the more documenta-tional motivation of explorers versus the subversive nature of the exploration sometimes performed by Garrett. This debate serves to highlight the complexity of the very idea of subcultures – they are forever in a state of becoming, which necessarily entails conflicting meanings, juxtaposed reasoning and internal schisms. But what this debate also shows is that once subcultures can be identi-fied, once they have begun 'concretising public interest' (Garrett 2014: 10) and attempt to 'put markers in the sand' as to their politics, they then engender their own systems of capture, their own hegemonic politics. These politics can often be very specific to a particular regime or locality. The next chapter will detail the Long Live South Bank campaign in more detail, but it was a specific reaction from the skateboarding community to actively resist the threat they faced from eviction from a very small but important locality on London's South Bank. Other examples include specific street art campaigns against particular movements such as the anti-Tesco campaign in Stokes Croft in Bristol previously discussed in Chapter 3 (see Figure 3.3). Street artists all over the world will have anti-establishment messages and iconography in their work. Some of it can be highly geopolitically charged, like the street art of Rex Dingler, an artist from New Orleans who used stencils to spray the phrase 'Somewhere in this city, this blood is real' in a few select places in Tel Aviv (see Figure 6.4). Various Tel Aviv officials told me that this campaign was a direct critical humanitarian commen-tary on the continual conflict in the region.

Some campaigns are more socially charged. For example, since 1983, the Lawson Street Bridge in the Redfern area of Sydney has been a site of social street art, with murals depicting historical, social and cultural messages (see Figure 6.5). Constructed collaboratively and maintained by residents of Redfern, the murals (on both sides of the street) are attempts to interject the contemporary city's aesthetics with non-consumption narratives.

Street art is often the subculture that generates the most visible and lasting resistive politics (although it is arguable as to whether street art can be considered *sub*cultural at all). It is, however, an identifiable 'thing' which can be utilised to express an agenda and feed into the construction, development and maintenance of a 'structure'. That maybe one of continual reproduction of the Creative City (see the next section) or one of anti-hegemonic activism, but in both cases, there are functionalities of objects that have obscured the realisation of their functions. A street that uses a bridge wall to make a humanitarian message rather than commercial advertising is no less adhering to a system of Baudrillardian signs than commercialist functionalities are.

This is not to say that anti-hegemonic, humanitarian, critical geopolitical or activist subcultural activity is somehow commensurate to the injustices and inequalities of the Creative City – far from it. Such activity espouses critique, discussion and debates and awakens urbanites to narratives that they otherwise

Figure 6.4 A Rex Dingler street art piece in Tel Aviv.

Source: Author's photo, 10 April 2011.

Figure 6.5 'Use condoms' creativity as a social act, street art on a Bridge near Redfern
        Station, Sydney.

Source: Author's photo, 17 April 2013.

would not be exposed to. Creative subcultural activity therefore provides a funda-mental counterbalance to the dominant voices in the city. They are contemporary instances of unitary urbanism that allow us to take stock of the constant bombard-ment of the spectacle and listen to the heteroglossia of the city. Creative subcul-tures, by subverting capitalistic functionalities of urban objects, are adding different voices to the urban topography. They are added layers of messages to the urban palimpsest. They are encouraging us, however briefly and fleetingly, to think a different city, one that encourages active participation and citizenship and resists passive consumption. But the creation of new functionalities, however counter-hegemonic they maybe initially, as the previous chapter indicated, risks engendering similar problems of heteronormativity and ablism inherent in the Creative City. Urban exploration, while espousing a more inclusive city, counter-ing the prevailing discourse and encouraging the 'redemocratisation' of public space, has been critiqued as being a highly gendered subculture. Mott and Roberts (2014) have argued the work of urban exploration is highly masculinised. They argue that:

> [The] geographic literature on urban exploration largely mirrors dominant presentations of exploring bodies found in wider popular urbex discourses. Specifically, we are struck by the authority granted particular exploring bodies: those performing an able-bodied, heteronormative and typically white masculinity. (Mott and Roberts 2014: 6)

The accusation of urban exploration as being masculinised and ablist can be considered an extension of the critiques of unitary urbanism. By celebrating the emancipatory capabilities of creative subcultures, we are normalising their inher-ent limitations of access within that celebration. Mott and Roberts argue that there is a paradox within urban exploration; if it emphasises a radical politics of resistance but portrays this through physical strength, fearlessness and heroism, then it is marginalising othered bodies just as systematically as the Creative City does. In effect, urban exploration, by becoming a subculture (through the subcul-turalisation process outlined so far in this chapter), has created similar injustices and inequalities that exist within the very system it is attempting to rebuke. Mott and Roberts (2014) note that urban exploration in fact intensifies the gendered nature of urban space as it deliberately seeks out those places that are already highly gendered (construction sites, sewers, etc.). Women, they argue, simply do not have the same access to these places. Also, people confined to wheelchairs are unlikely to be able to partake in urban exploration. Other physical and mental disabilities will also act as barriers to participation. Such a critique could quite easily (and with justification) be levelled at parkour, skateboarding, buildering, street art and others.

Urban exploration is a creative subculture as it is subverting the functionality of the urban places and sites that it connects with. Those who infiltrate a building site or go into a sewer are eschewing the capitalistic functionality of those objects and realising an alternative function that expresses the desire to create a new way of thinking, new histories and alternative subjectivities. But because it also has the potential to reproduce the same inequalities that we saw in the Creative City, then it is easy to see why the creation of subculture is a critical point. By the time we can call something a 'subculture' it has already built up systems, signs, rules, communities, frameworks and guidelines that have as much exclusionary potential as they do inclusionary and emancipatory powers. They have stopped *becoming* and have become *being*. And, it is this 'state' of being that Mott and Roberts (2014) are countering, and one that should be rebuked in the creative city. Garrett and Hawkins (2013), in responding directly to Mott and Roberts' (2014) critique of urban exploration, argue that it is a practice that comes about through an affective corporeal politics that is far more heterogeneous and multiple. Instead, they propose that we should 'ask how it is that these heterogeneous matters and forces come to compose durabilities of orderings such as race and gender' (Garrett and Hawkins 2013: 16). In other words, urban exploration is the *durable* or visible manifestation of a whole range of multifarious happenings, an assemblage of corporeal potentialities that, at its most concrete, espouses gendered, racial and normative biases. However, these biases are inherent to all creative subcultures. They are all simply more durable and hence more visible beings of the subculturalisation process. In the previous chapter, I argued that by becoming-minor, by engaging in lines of flight, is how we can realise a creative city free from the injustices and inequalities of the Creative City. The problems of heteronormativity (and the other characteristics that can be rightly critiqued in subcultures) come about when we stop engaging in flight, when we stop attempting to become minor. Therefore, in order to maintain the creative city, we must continue to flee, to always look for new possible ways to realise function to satisfy desire-production. For if we stop and begin to contextualise our endeavours (by creating subcultures), it is then that all the issues and problems can begin to re-emerge. Yet in countering these problems, there is another way in which urban subversion can occur. The evolution, tensions and schisms of creative subcultures come about through the constant attempts to defend against such biases. There is messiness to their makeup, articulated by urban exploration's multiple narratives, but also by the relative tensions within the skateboarding and parkour communities (see next chapter). Such conflicting, fluid and complex processes can be thought of as the subculture contorting to the pressures that such heteronormative biases can create. Some of these tensions will also be due to internal hierarchical politics and infighting, and can represent the internal competitiveness that can inculcate further mirroring of Creative City processes. But urban subversion can further be thought of as internal struggles *within* urban creative subcultures to renegotiate the inherent problems that can formulate through their continued ossification and identification. In other words, there is becoming-minor within subcultures.

It is worth reiterating that such problems (such as the heteronormativity of urban exploration that Mott and Roberts (2014) were critiquing) are relative to the far more wide-ranging and systemic problems of the Creative City. It is, to some extent, engaging in a Freudian narcissism of small differences when compared to the overarching injustices of the Creative City that were detailed in Part I of this book. Urban exploration, despite its critiques, far from entrenching masculinity in the subversive politics of urban access, can be used to inspire people to engage in activities that are less aggressively reactionary but in no way any less creative. The state of *being* that many subcultures find themselves in does reproduce inequalities, this much is clear. Yet they can also act as pointers for others to engage creatively with their city. They can act as catalysts for desire-production of other functions throughout the urban terrain. They can show us that other voices, other layers, other functions and other possibilities exist. Moreover, they recognise the inherent biases and attempt (albeit some more successful that others) to redress them. Overall, they show us that other cities are possible. This, I argue, is a far more desirable outcome than a subculture pointing *directly back* into the Creative City.

This is because we need to be aware that identifiable subcultures have in themselves the potential to further the development of (and therefore exasperate the problems of) the Creative City through their co-option into it, rather than maintaining an opposition to it. Accusations of selling out plague subcultural practice, but it is often far more complex than this. 'Selling out' has more to do with a symptom of the Creative City appropriating subcultural activity for its own development rather than individuals looking to profit from their activities – it is part of the systemic need of the Creative City, the constant search for novelty. Chapter 2 discussed how neoliberal practices are now thought of as complex assemblages that are constantly searching for novel ideas, politics and practices that can be used to generate wealth. The Creative City, as Part I show-cased, is the latest, refined iteration of this need for novelty, and so urban subcultures are viewed as practices that can become profitable. Yes, individuals can be said to 'sell out', but it needs to be understood as part of a much more complex assemblage of Creative City development. Hence, the next section will show how particular subcultural activities can be subsumed into the Creative City.

Selling out?

The 2010 film, *Exit Through the Gift Shop*, is a 'documentary' (some have refuted this, however, arguing the film is all an elaborate hoax) about a man called Thierry Guetta. He runs a clothes shop in Los Angeles but has an obsession to film everything – he films his food, his children, his friends and his surroundings. His camera is always with him and always on. His cousin is a well known street artist called Invader and Guetta begins following him around during his nocturnal, illicit street art activities. Guetta states that he wants to start making a documentary about street art and gets his cousin to let him accompany and film many other famous street artists as they go about their work. After some time, Banksy, perhaps the world's most famous (or infamous) street artist, visits Los Angeles, and Guetta begins to accompany him too. In 2006, Banksy asks Guetta to document a show he is putting on in Los Angeles called 'Barely Legal'.

This show was shrouded in secrecy, but upon closing after the three days it was visited by huge numbers of people including celebrities and institutionalised art critics. After the event, street art became much sought after:

> Barely Legal marked the point at which street art was forced into the spotlight, attracting sudden interest from the art establishment. In the months that followed [the show], prices for work by leading street artists rocketed with collectors rushing to get in on this exciting new market. Street art had become a white hot commodity.
>
> (Narrator, *Exit Through the Gift Shop*: 50m 15s)

Guetta was then challenged by Banksy to finally make the street art documentary. It does not turn out too well, and so Banksy attempts to make the film himself. In order to stop the now rather annoying Guetta from intervening, Banksy tells him to go back to Los Angeles and 'put up more of your posters and make some art, and have a little show' (Banksy, *Exit Through the Gift Shop*, 56m 2s). Guetta takes this task to heart and starts embarking upon a street art campaign, calling himself 'Mr. Brainwash'. He fervently goes about plastering Los Angeles with his stencils and pre-printed posters. After a short time, though, he goes about trying to put on an exhibition. He employs an entire team of graphic designers to reproduce hundreds of prints that 'riff' on Banksy's work (perhaps the politest way of putting it). In a very short space of time, Guetta has publicised the event (with the help of a quote from Banksy) and produced over 200 works. By the end of the opening week, he has sold nearly $1 million worth of art. The film suggests that Guetta had become art's 'next big thing' in a very short space of time.

Debate in the critical response to the film has speculated that the entire film was staged. Banksy and Guetta refute this, but really, whether or not the film was staged (whether it is indeed what Catsoulis (2010) called a 'prankumentary') is irrelevant. In either scenario, it points toward how the 'subculture' of street art has become intertwined into the neoliberal assemblage of the Creative City. The film often critiques the 'art world' (although never properly defines what this is) as being duped by Guetta. Given that he spent years following and filming street artists, he was able to recreate a formula for financial success relatively quickly. He knew the styles of art, the techniques used (in terms of producing the art and its clandestine distribution) and the exact way to publicise and 'spin' it to a mainstream consumer audience. By identifying (and rather brazenly copying) the prevailing trends and knowing the 'boundaries' of street art, Guetta was able to financialise the subculture in a rather spectacular way. Moreover, he completely industrialised the production of street art. He had a factory and a workforce that produced these artworks in a Fordist-style mode of production. It is very much the industrialisation of a creative subculture.

There is a vast array of literature that recounts the history of graffiti as well as its relationship with street art (in particular see Cresswell 1996; MacDonald 2013; Waclawek 2011) but there is a general consensus that the crudest of boundaries can be identified through legality: street art is legal graffiti. The 'modern' style of graffiti was synonymous with 1970s New York, where subways trains acted as canvases for increasingly more elaborate and colourful 'taggings' (text and symbols that convey someone's name, a message or simply an esoteric word, often a nickname). Snyder (2009) notes how it was in 1972 that the tagging of trains became widespread after an 'innovation' in the way in which spray cans could be used to cover a wider area. Soon whole train carriages were being adorned with large colourful, 3D-style text, mostly at night to avoid capture. The style of the text became very much part of New York's cultural landscape and until very recently one could visit 5pointz in Long Island in Queens, New York, to see highly visible and intricate examples of this style (5pointz will be touched upon further in the following chapter). In the years since, graffiti has spread around the world and it is now difficult to take a journey in any city without coming across some form of recognisable graffiti. Using the articulation of creativity above, it is fairly clear that the creativity of graffiti comes from how an alterative function of the side of a train or a blank wall is realised by the desire of a 'graffiti artist' to express a particular emotion, belief, opinion or affective experience. The functionality of the train is its part in urban mobility, as a conduit for passengers to get from one point of the city to the other. Its part in the system of urban governance is to aid flow, to be an arterial route in a city of order. The train's universe of functions is therefore masked, streamlined by the intensity of urban mobility and flow of people to work and back. However, in the 1970s, the 'taggers' realised another function based upon their unsystematised desire. The train became functional as a canvass to express a range of affective emotions. The smog of functionality had been cleared for that briefest of instances. They were of course marginalised by New York City, they were arrested, deterred and 'othered', which is how we saw the initiation of the subculture of graffiti; there were groups of people involved, united by a common grievance and a similar marginalisation practice. They then started encouraging each other, sharing ideas, collaborating and even competing with each other (Snyder 2009); this socialisation continued the *subculturalisation* of graffiti. But the subculture also created rules, frameworks and guidelines. Moreover, the very motivations for tagging trains ossified from that initial spark of desire-production; they formulated their own functionalities that were part of a system not of capitalistic urban development, but of resistance, subversion and activism.

So graffiti is a *true* creative act, but the subculture that grew from it made graffiti an identifiable 'thing' and the creativity begins to be crowded out by the layers of functionality. Subsequently street art has become a major part of the art world, since collectors starting paying millions of dollars for works by former train taggers. Hence, through the identikit work of Banksy, Mr. Brainwash and others, graffiti has stopped being truly creative and become creative in the other dogmatic sense of the word (Dickens 2008). It adheres to a system of creativity that has aims, goals, frameworks and rules (usually orientated toward profit-making). Given that many 'street artists' have spent their formative years being marginalised by urban governance, they have been subculturalised and hence have been using their skills as a functionality of subcultural creativity. In other words, the realisation of alternative possible functions (of that train or blank wall or any graffitied urban object) has been achieved and is now part of a functionality of graffiti as a subculture. As a result, that functionality can be drawn into the Creative City. The system of signs that the functionality of an object feeds into is what the Creative City co-opts and reappropriates. As it is a neoliberalised assemblage of governance, it has the systemic requirement for novelty, for constant tweaks and additions in order to maintain development and all the wealth and profit that it generates. The Debordian *spectacle* of urban consumption of creativity 'further ensures the permanent presence of [the] justification' (Debord 1967: 6) for a system of control. But such control, such addition and tweaking of the Creative City cannot happen without a system of signs, without a framework for justification. This is why pure acts of desire-production, becoming-minor, sparks of true creativity that cut through functionality are not 'capturable' by the Creative City because they are not recognised. Quite simply, they cannot be *seen* by it. By adhering to a system, creative subcultures become recognisable, articulable and therefore potentially profitable. Individuals that practice a subculture have agency in this process – Guetta, for example, through his actions of industrialising 'street art' production, eradicated any function and creative desire from the city and instead formulated a functionality that was very much part of the 'art scene'. His actions, though, are part of the sociability of the subculture of street art. (It is questionable of course to argue whether it can be considered a subculture at all now – the appearance of Banksy works recreated in Lego is perhaps another indicator (WebUrbanist 2014)). The intense collaborative and competitive actions of the multitude of street artists around the world (a good number of which Guetta documented) catalyses the systemisation of street art. The very subculturalisation process that creates an identifiable subculture also creates its ability to be co-opted. 'Selling out' therefore is not an individual deciding to make money from his or her art – the co-option has started long before that. It is collective social action that creates subcultures, but in so doing creates the systems of its own (potential) appropriation into the Creative City.

Put another way, the identification and ossification of subcultural boundaries and particular trends/fashions allows individuals, as members of a particular subculture, to be targeted by advertisers for profit. To use a different example to street art, particular items of clothing have become popular among skateboarders, enabling the introduction and eventual prevalence of particular clothing brands. In parkour, there are certain dress codes that prevail and hence these can be styled, branded and sold to traceurs. We also see the appearance of 'heroic' individuals within particular subcultures who are often sponsored by the major brands and companies with a vested interest in selling clothing or items within the subculture. For example, Tony Hawk, a skateboarding 'superstar', is sponsored by Quicksilver who make skateboarding apparel and paraphernalia, and Sebastian Foucan, one of the founders of parkour, has been sponsored by Nike, often appears in advertisements for them and has appeared in Hollywood blockbusters. Individuals who become synonymous with a subcultural practice are not somehow lone pioneers blazing a trail of unconformity that champions an alternative to the hegemonic urban condition. This is merely a 'system of signs' that is applied by the apparatus of capture as they need to be viewed as new, alternative and distinct if their full profit-potential is to be realised. It is the Debordian spectacle in action. However, these individuals' agency is founded upon an already subculturalised process which has seen skateboarding, parkour, street art, etc., become some*thing* that can be subsumed into the Creative City. The creativity of these subcultures has long been lost to functionality, and is being recoded into the dogmatic creativity that can be replicated internationally.

So, subcultures are rife for co-option. They are the fertile ground for the Creative City to plant new seeds of commercialism. As subcultures grow, become more recognised (digital and virtual technologies have sped this process up exponentially), have more rules, guidelines and how-to guides (see the yarn-bombing self-professed 'DIY manual' (Moore and Prain 2009) for example), they become more susceptible to profiteering. Selling out is simply the final 'piece' of this process: the financial reward to specific (more entrepreneurial) individuals for allowing the Creative City to reappropriate a subculture. What we are seeing more recently though is the speeding up of this process. From the moment of initial creativity, the realisation of an alternative function of an urban object to the full co-option of the subculture into the Creative City is happening quicker and quicker. Sebastian Foucan and David Belle are said to have been two of the 'founders' of parkour when they were children in the late 1980s (Angel 2011). Today, it is a globally recognised phenomena that has appeared on TV and film, and 2009 saw the World Championships in London (sponsored by Barclaycard of course). It is important to note that the transition of parkour from an activity that urban officials would marginalise to a commercial activity is not a linear process, and the presence of Barclaycard as a sponsor of the championships is just one instance (albeit a rather crass one) of financialisation. Indeed, within any identifiable subculture, there will still be individuals who continue to be marginalised, while others will be gaining a financial reward for their subcultural 'creativity'. The 'phasing' of the processes I have outlined so far in this chapter is not sequential. They form part of atemporal, complex and multiplicitous happenings. Indeed it is important to note that even the relatively stable concept of the subculture that I have alluded to throughout this chapter (a la Fischer 1975) is porous, ever-changing and always in flux. This is why, again, it is important to adhere to a Deleuzian notion of becoming-minor, to resist a state of *being* with all the normalising biases that it entails. Engaging in lines of flight means always uncoupling from majority rule, seeking out new ways of becoming that are alternative to the apparatus of capture. And so it is here that we have come full circle: an identifiable subculture is as much a system of instrumentalism as the Creative City. A subculture can be equally as hegemonic as it adheres to a system of signs that is constructed by functionalities. A subculture bears all the characteristics of the Creative City, just waiting to be appropriated.

Summary

At the beginning of this chapter I asked, what is 'urban subversion'? Through a discussion of the act of creation and how it sparks the subculturalisation process, we have seen that urban subversions are more than merely a pseudonym for urban creative subcultures. Urban subversion is being complicit in the idea that we can flee the apparatus of capture and subvert the hegemonic city. It is the initial spark of desire-production, the need for socialised collaboration and the desire to subvert the Creative City to espouse more social justice combined. It is the desire to not accept creative subcultures' inherent biases and to redress them while maintaining subversive politics. Urban subversions are not individual subcultures, they are not just those groups of people who are subverting the dominant ideology, they are not ossified groupings that can be labelled. They do not replicate the heteronormative, ablist and class biases of existing power structures. They do not perpetuate social injustices for the sake of resisting hegemonic power. Urban subversion is not a noun. Urban subversion is a process: it is the partaking in practices that simply escape majority urban creative thinking. Urban subversion is the practice that allows us to realise a creative city.

I have shown in this chapter how, through using ontologies developed by Baudrillard and Delueze and Guattari, we can epistemologically understand how a creative city can be thought of as ontologically different to the Creative City. The creative city incorporates urban subversion and in many ways is the vision of urban *being* that we can aspire to. But it is always critical to note that by acting in this way, the temptation is to cease becoming, to cease urban subversion and to become a state of being. This, as we have seen, risks mirroring the inequalities of the Creative City or, worse, becoming appropriated by it. We need to continually engage in lines of flight. There is one more ideological trope which allows us to grasp more fully these ideas. De Certeau (1984) in *The Practice of Everyday Life* articulates how 'tactics' within the everyday have an innate power to react, to resist and to reclaim. The word 'tactic' implies a less harsh oppositional stance to hegemony, rather a 'softer' subversion that does not reject or transform existing power structures but metamorphoses them to 'function in another register' (De Certeau, 1984: 32). He distinguishes between what he calls 'strategies' and 'tactics'. Strategies are much like apparatuses of capture in that they contain, compartmentalise and co-opt; they 'assume a place that can be circumscribed' (1984: xix), or a 'place that can be delimited as its own' (1984: 36). De Certeau offers some examples of such strategies, including a business, an army, a scientific institution and, critically for the arguments of this book, a city. Tactics, he argues, are those instances of incursion into a strategy:

> The place of the tactic belongs to the other. A tactic insinuates itself into the other's place, fragmentarily, without taking over in its entirety, without being able to keep it at a distance. It has as its disposal no base where it can capitalize on its advantages, prepare its expansions, and secure independence with respect to circumstances … Whatever it wins, it does not keep. It must

constantly manipulate events in order to turn them into 'opportunities'. (De Certeau 1984: xix)

So, a tactic infiltrates the totality, subverting it from within, while never claiming a functional or identifiable space. The language here is very much aligned with the reactionary ethos of some subcultures (such as urban exploration) precisely because, as has been noted, it is important to maintain such politics as they enrich the heteroglossic voices of the city. It can utilise the existing infrastructures on offer (from the strategy), but never claim a space or a territory of its own as this would be to engage with the system on its own terms, i.e. creating oppositional structures and replicating their inequalities. The aspatiality of the tactic renders it elusive to appropriation by strategic control, as strategies occur through the claiming of space. Therefore:

> This nowhere gives a tactic mobility, to be sure, but a mobility that accepts the chance offerings of the moment, and seize on the wing the possibilities that offer themselves at any given moment. It must vigilantly make use of the cracks that particular conjunctions open in the surveillance of the proprietary powers. It poaches in them. It creates surprises in them. It can be where it is least expected. It is a guileful ruse. (De Certeau 1984: 37)

I want to offer a brief example of such a tactic. In 1995, a Russian-built T34 tank arrived in London (apparently for use as a prop in a film). Afterwards, a property developer named Russell Gray (who lives in Southwark) purchased the tank for a rather ostentatious but very *tactic*al (in the De Certeauian sense) reason. He had been refused planning permission to build a house on a particular plot on Mandela Way in Southwark. In response, he applied to put a 'tank' on the property instead. The council understood this to mean a water/waste tank but Grey meant a very different kind of tank. Permission was granted for a tank and now the T34 sits proudly on this site (see Figure 6.6).

The tank is now regularly decorated in different patterns and is a permanent feature of the landscape. It is a De Certeauian tactic because Gray subverted the strategy from within. He used the existing structures of planning procedures, legal systems and the ambiguities of language to create something which exists within the system but very much snubs it. As a final act of subversion, the tank's weapon is pointed toward Southwark Council's head offices.

These kind of activities can *take place* in that they reappropriate, reconfigure and enliven a particular place through existing structures. But as soon as these practices take the *space* of the city through engaging with the inherent strategies (neoliberal urban development for instance), then they cease to be *tactic*al, and become part of the city's strategy. If this tank was to be used to advertise products, if it began to act as a place-making feature, if it was used by Southwark Council to somehow catalyse the development of the area, it ceases to be tactical and becomes part of the strategy of the Creative City (none of these things have happened ... yet). Similarly, if its subversive politics were such that it became

Figure 6.6 Southwark's T34-tank, London.
Source: Author's photo, 6 May 2014.

gendered, ablist or somehow exclusionary then it has begun to replicate certain inequalities. But by infiltrating or intervening in the city in a tactical way, by taking place momentarily but never claiming space, these moments of urban subversion are perpetually in motion, which engenders a purer politics of transgression (Cresswell 1996) rather than complicating the contextualisation with an binary inflection of 'us' versus 'them'. The tank is not in motion, but it is regularly decorated differently to make sure that its tactical-ness remains just that little bit in flux. A tactic is the moment of release from the urban strategy, the realisation of function, the initial spark of creativity. Once these activities accumulate, once they form a broader politics (of resistance to, or amalgamation into, the Creative City), they cease to be tactics and become something else. As De Certeau (1984: xix) so succinctly put it (and is hence worth reiterating), 'whatever it wins, it does not keep'.

This chapter has shown that urban subversion helps us to realise a creative city. It has detailed that creative urban subcultures have the power to catalyse a more open, experiential and democratic city, but they also have the ability to increase inequalities too through co-option into the Creative City. Such a process needs to be guarded against if we are to continue to engage in urban subversion.

Luckily, we have help. We can recognise the appropriation of subcultures with one critical aspect – that of place. In the same way that the Creative City policy *demands* real estate developments that aid its dispersal and development (Cultural Quarters, Media Cities and so on), the zoning of subcultures has the same effect. If the Creative City can create manicured places where subcultural activity can proliferate, then they are well on the way to profiting from them and appropriating them completely. The *placing* of subversion is very important to the Creative City. So we need to recognise the systems, processes and strategies that attempt to 'zone' subversion if we are to continue our attempts to realise a creative city. And it is to these that the next chapter will turn.

7 The places of subversion

London Waterloo station is a landing spot for the city. The commuter trains that shuttle in and out of the transport hub carry millions of people every day to and from the South of England to its capital city. It has the largest number of platforms of any station in the UK, and until 2008, it was the gateway to continental Europe via the Eurostar terminal. It is also one of the main termini for the South Bank, an area of London that stretches along the South side of the River Thames from London Bridge to the East, through Bankside and ends somewhere near Westminster Bridge. The area incorporates many tourist attractions including the London Aquarium, the Southbank Centre, the Tate Modern and the iconic London Eye. The area has over the years become notably more developed, with new gardens, attractions and leisure and retail units being added every year. Being adjacent to one of the main transport 'gateways' into London, it benefits from large visitor numbers and is a classic 'honeypot' site.

The area has been a major focus of much of my attention on urban subversion. This is because as well as being a major tourist destination, it is also the site of a number of 'iconic' creative and subversive subcultural places. These range from 'dead spaces' that have been reappropriated for subcultural use, sites that have been (not so) subtly coerced into housing subversive subcultural activity, private commercial spaces that are frequented illicitly, and also some places nearby where subcultures gather to practise, meet up or are regular 'spots' of subversion. Indeed, over the years, a geographical 'sliver of subversive spaces' (to give it a crass, alliterative label that can be used as shorthand) can be identified that runs from the undercroft at the Southbank Centre where skateboarders ply their trade, to the Battersea Power Station which is one of the most visualised, written-about and fetishised locations in the urban exploration community (Garrett 2013). The range of subcultural uses of this area is not unique – there are many other spaces around London where subcultures cluster. There are even spaces in other cities that no doubt house a similar range of activities if not more so. However, in terms of their relation to the debates that are central to this book, namely the realisation of a creative city from the rubble of the Creative City, I would argue there is no more an appropriate suite of places. Why? Because not only has London enacted a number of Creative City policies, it is a Global City par excellence (see section 2.2). It is a city that is rapidly gentrifying and hyper-gentrifying, has major housing crises, over-inflated prices and more billionaires than any other city in the world (Wood 2014). It also contains the overwhelming majority of the UK's cultural funding and facilities and houses a multicultural, multi-ethnic and multiplicitous population. Because of this, London has become seemingly very efficient at co-opting new creative practices. (The recent debate in the mainstream media about 'Shoreditchification' is a prime example (Proud 2014)). As will become apparent with a discussion of the undercroft, the Vauxhall Walls and Leake Street, there are a myriad of institutions, forces, ideologies, affects and assemblages that create tensions in these places between the people who practise creative subcultures in these places, and the overarching forces of the Creative City.

The undercroft

The Southbank Centre in London has over the years been a cultural hub for the city. Initiated for the Festival of Britain in 1951, it is now the site of a number of important cultural institutions such as the Royal Festival Hall and the British Film Institute; it is a hive of cultural activity and a popular tourist spot. The (in)famous brutalist architecture that pervades the buildings, walkways and bridges, rather than creating an intimidating atmosphere of enclosure, striation and forbearing, instead has become somewhat of an ostentatious statement of its incommensurability with the surrounding urban aesthetics, as well as its cultural provisions. In other words, the brutalism that stands in stark contrast to the neoclassicism on the other side of the river and the surrounding glass monoliths of real estate development is not 'out of place' but rather pervades a sense of comfort that such architectural awkwardness belies. For while the main occupation and activity on the site has been the 'high-class' cultural provisioning of the Haywood Gallery, the Queen Elizabeth Hall and the Royal Festival Hall, the exterior spaces have, over the years, become subaltern, with a vast array of street performers, market stalls and of course, the skate spot at the undercroft. Such 'alternative' forms of cultural provisioning, however, have slowly been eroded through incessant commercialisation, commodification and control of the 'public' space towards those activities that fall under the rubric of the creativity paradigm. Particularly within the last decade or so, the South Bank, under the increased privatised and corporatised characteristics of the business-like management agency and landlord, has taken on consumption-orientated aesthetics. More chain restaurants and cafes have opened, generic public events have been staged and sponsored by large corporations, and the area has seen farmer's markets, boutique artisan 'pop-up' stalls and professional street performers proliferate. And in a move that aimed to drastically change the architectural style, urban aesthetics and no doubt the level of overt management and securitisation of the area, the South Bank Centre in March 2013 unveiled the 'Festival Wing' plan. The initial plans (which have been changed over the course of time, but at the time of writing, are still very much in the offing) included a complete architectural overhaul of the Festival Hall, which would have cloaked the building's brutalist pervasiveness with generic contemporary building materials such as glass, and reconfigured many of the resident businesses, communities and cultures into a meta-narrative of culture-led development. The Southbank Centre are determined to develop the area as 'London's next Cultural Quarter'. Therefore these plans can very much be considered further examples of what was discussed in Chapter 4 – the physical (and therefore most profit-making) manifestation of the Creative City policy process. The claim is that such a development, which is being undertaken with corporate sponsorship, is to 'realise its [the South Bank] vision to deliver a larger and more ambitious arts, educational and cultural programme across the site for all its visitors to enjoy' (Southbank Centre, 2013a: 1). Part of the redevelopment of the Royal Festival Hall area initially included 'the under-used spaces from the undercrofts' (ibid.) being turned into retail outlets. Moreover, the plans involved building a £1m skatepark 150m further West under Hungerford Bridge, which was to be a 'new riverside area for urban arts' (ibid.) One of the key messages that has been used with the marketing of the plans has been 'Want more art for more people?' (see Figure 7.1) and 'Culture for all'.

The 'under-used' space the plans referred to (sometimes called SLOAP – 'spaces left over after planning') that had been earmarked for retail outlets is currently a spot where skateboarders congregate, informally referred to as the 'undercroft' or the South Bank skate spot (see Figure 7.2).

Figure 7.1 A 'Want more art for more people?' Poster in the Southbank Centre.
Source: Author's photo, 12 October 2013.

Figure 7.2 The undercroft, South Bank, London.
Source: Author's photos, May and October 2013.

From my intense discussions with skateboarders at the site and general involvement in the community, people have been using this place informally since 1973 when people first started to sneak in at night with their skateboards and use the area to hone skills, practise tricks and generally enjoy the community spirit of skateboarding (although some sources say it was first used for skating even earlier in the late 1960s). From then, the undercroft has, over the years, continued to attract skateboarders from all over the world. Its undulating surface, small concrete protrusions, smooth concrete textures and the fact that it is protected from the elements means that it is a 'perfect' spot for skateboarders. While initially the landlords attempted to marginalise and displace the skaters through increased security patrols and fines, and latterly using the area deliberately as a dumping ground for gravel and general construction detritus, today the spot is a major feature of the South Bank and they are 'allowed' to perform their skating. The area now has BMXers, trial-riders and graffiti artists and is home to a whole range of creative subcultural activity. There are of course internal conflicts as there are with any collective of subaltern practitioners, but for sometime now the area has been an integrated part of the South Bank's (sub)cultural ethos. Tourists flock to it and stop and take photographs, it has been used to stage music videos, it has been used as the backdrop for magazine shoots (see Figure 7.3) and it even featured in a Tony Hawks video game. Coupled with some of the other perhaps less mainstream, certainly less commercialised forms of cultural consumption (such as the second-hand book shop under Waterloo Bridge), the skaters of

Figure 7.3 Commercial use of the undercroft skate park.
Source: Author's photo, February 2013.

the undercroft have contributed to the 'subaltern' spirit of the South Bank over time, but now, represent the last bastion of such activity, surrounded by an ever-increasing wave of commercialisation and private, micro-management of public space.

What makes the area of particular interest to the ideas of reformulating a creative city is that the skateboarders originally reappropriated the undercroft for their own use. They saw the area as a perfect arena in which to practise their skill and craft, and not as a place that serves a particular function (which is architecturally and pragmatically at least to support the walkway and Festival Hall above it). Initially the skaters were chased down, marginalised, told that they were not allowed to be there. This is very common in the skateboarding community. As Borden (2001) noted, skateboarders in the 'early years' (1960s and 1970s) sought out empty swimming pools in the Hollywood Hills as they provided the perfect 'curved transition from base to wall' (ibid.: 33) and emulated the movement of surfing (often seen as the mother of skateboarding). Sometimes these pools were deliberately drained by their owners, but as the practice spread, skateboarders began looking for and skating in more and more empty residential and public pools, whether they had permission or not. This act of using empty swimming pools in a way in which they were not originally designed is yet another instance of people realising an alternative function of an object, subverting its functionality (as an item of leisure and ostenta-tious habitation) and connecting viscerally with that object to express desire-production. The practice of such reappropriation has been argued as having displacing tendencies of its own. Skaters who target particular spots can displace homeless people who currently use those spaces precisely because they are disused and unpoliced. As such, skateboarders have been viewed as the 'shock troops of gentrification' (Howells 2005), but such an accusation speaks more of the processes of neoliberal development that actively seeks places that already have an element of use and an 'identifiable' way in which profit can later be extracted. The reuse of the undercroft can be seen in such a way. Moreover, the continual usage of the undercroft in this way for over 40 years has created an identifiable, coagulated place of urban subcultural activity, of urban subversion made concrete. It repre-sents the place-based intensification of subculturalisation detailed in the previous chapter. The complexities of the place of the undercroft (i.e. its heterogeneous makeup) point to another way in which we see the ossification of skateboarding as a subculture. It has been called the 'home' of UK skateboarding (much like Love Park in Philadelphia was considered the 'home' of US skateboarding before it became overly policed), and is *the place* to go if you are at all invested in the subculture. I am not a skateboarder but in my youth I frequented the spot, often with family and friends and watched them using the space. I even dabbled in it myself before injury, embarrassment and a distinct lack of ability brought about a swift end. Over the course of the 40 years people have been skating the undercroft, it has woven itself into the identification of the subculture. Its visual iconography has become synonymous with skateboarding and as such is an integral part of the identification of the subculture. From initial attempts at direct marginalisation to the social grouping and collaborative activity that brought them back, and now the general 'acceptance' of it as an important place in the skateboarding subculture the place-based characteristics of the undercroft are intertwined into the subculturalisa-tion process. So it is easy to understand the uproar, dismay and general disbelief within the community when the Southbank Centre announced that it was to be expunged. And given the counter-cultural, resistive nature of skateboarders (Borden 2001), they were not the kind of people to not try and resist this.

'Long Live Southbank' (LLSB) is the campaign institution that was set up in response to the plans for the Festival Wing. The campaign collected over 150,000 signatures, created films and art exhibitions and continually attempted to enter into dialogue with the Southbank Centre (SBC – the landlords who are spear-heading the redevelopment plans). They also managed to deliver over 40,000 objections to Lambeth Council over two events, making it the largest ever objec-tion to any one planning application in UK history. The main thrust of LLSB's argument is optimised in their tagline 'Preservation, not relocation' and 'You can't move history'. This clearly denotes a topophilia, a love and longing for the place that has been created through decades of subcultural reappropriation. The reaction to being forced to move to a new area for the urban arts, albeit only 150m away at Hungerford Bridge, is a fundamental and visceral response to changes imposed by urban hierarchies.

The skateboarding community have 'made' the skate spot what it is today through years of defiance and counter-cultural activity, and when threatened the response is to create a politics of anti-hegemony, but a politics that is firmly about the urban *place* that is the subject of appropriation. Like urban exploration discussed in the previous chapter, LLSB has formed a creative subcultural community that is in stark opposition to the cultural and urban hegemony (the Southbank Centre and the Creative City rhetoric of which it is intrinsically a part). And therefore also like urban exploration, it risks duplicating the very inequalities of the Creative City that subcultures attempt to rebuke. Indeed, skateboarding has similar masculine overtones and a distinct corporeality that eschews less-abled bodies. Skateboarding has been critiqued as being overtly the pastime of white, middle-class (Karsten and Pel 2000) young men (Atencio *et al.* 2009). Femininity in skateboarding is often cast as somehow 'inauthentic' given that women are often unable to perform the tricks and stunts with the veracity and ostentatiousness that men do (arguably, parkour has similar traits). This issue becomes particularly acute when place is factored in as the intense sociality of the undercroft can increase the pressure to 'perform'. It can risk replicating 'playground politics' where invariably masculinity and physical prowess dominate.

So skateboarding (or subculture more generally) contains normative biases brought about through the subculturalisation process. Marginalisation, sociality and eventual identification into a 'thing' means replicating inequality. However, to guard against such normalising tendencies we must realise the creativity inherent to the initial formulation of the skate spot in the first instance. This means a reiteration of its subversive and resistive politics, but also a celebration of the *true* acts of creativity that brought about this place-based subculturalisation process. The LLSB campaign that was created in resistance to the Festival Wing is a prime example of the way in which the politics of a subculture can begin to subvert not only the Creative City, but also the gender, racial and class biases that the subculture can manifest. However, unlike within the neoliberalised, capitalistic tendencies of the Creative City, such biases *themselves* are constantly resisted and subverted at the undercroft. As was discussed in the previous chapter, creative urban subcultures are subversive not just because they subvert functionality, but because they constantly attempt and have the desire to subvert the normalised biases that concretise over time. Much of the commentary from the SBC (including those that informally and confidentially spoke to me) argued that the skateboarders at the undercroft are nearly always white, middle-class males. However, having visited the site on many occasions during the campaign, this was far from the case. The undercroft has an ethos of inclusivity. While not explicitly codified in any particular guidelines, the skaters (those that were accomplished and those not so) adhere to an ethics of geniality, acceptance and non-abusiveness. While not always followed, those that run counter to this are often castigated. The space is inherently democratic. Moreover, the undercroft is one of the last few remaining spaces in the South Bank area of London where often marginalised young people (of various ethnicities) can come and not be harassed by the police, security guards or private land owners. Many of the skaters I spoke to were of school age and lived nearby in the lower-income areas. They said that the undercroft was one of the few places nearby they could play in safety. Other older users of the undercroft spoke to the area's community spirit, where people could go to learn from more proficient skaters. What is perhaps most telling is that the threatened destruction of the undercroft caused mass activism among a generally apolitical community. Many skaters told me that the LLSB campaign was the first time they had been involved in any political collective before. Previously they had seen politics as something that did not concern them, confined to the ornate building on the other side of the river. Yet, when hegemonic urban forces threaten *their* place, the undercroft, they suddenly became aware of the avenues of political action. They signed online petitions, wrote to their MPs, campaigned and generally began to care about the city around them. It awoke an activist mentality and enlivened people (some as young as seven or eight years old) to become active urban citizens. So the undercroft provides a space for cultural engagement, social inclusivity and healthy living, all the qualities (and more) that the SBC were championing with their tagline 'Culture for all'.

In January 2014, the Mayor of London Boris Johnson announced that while he supported the Festival Wing plan that the SBC was initiating, he said that it should not be developed at the expense of the undercroft. A week after this response, the Southbank Centre said that they would redesign the entire plan. In September 2014, 18 months after the initial Festival Wing plan was released, the undercroft's future was secured. A joint press release by LLSB and SBC said that any Festival Wing plans that go ahead in the future would not include the removal of the undercroft. This was heralded by many as a victory for LLSB, as it safeguarded the future of the undercroft. Of course, only time will tell. Given the neoliberal philosophies that are inherent in the Creative City policy discourse (of which the Southbank Centre is a part) it would be hardly surprising if further plans for development were perhaps more subtle but no less determined in their appropriation of the skate spot.

The 'Battle for the Undercroft' then exemplifies the process of *placing* subversive activity into the urban political economy by opposing it. To reiterate the arguments of De Certeau (1984), there is a sense that because the skateboarders are claiming an ownership of this place, they have become less tactical and more strategic. The subversive activity has, in effect, drawn 'battle lines' against the urban hegemonic forces (in this instance played by the SBC) and entered into a political dialogue with the city. This was exemplified with many blog posts, newspaper articles and opinion pieces on both sides of the argument, with accusations of misinformation, bullying and manipulation. The ultimate vindication of LLSB by the Mayor and their ultimate victory speaks to the huge groundswell of support that the campaign achieved. This was done through daily activist activity. The manning of a desk at the undercroft every day which obtained signatures for the petition from passers-by; online videos that spoke not only directly to the SBC to counter some of their spurious claims, but also that delved more deeply into the philosophical ideas behind the undercroft; artistic endeavours at the site and online; large-scale petition deliveries to the Lambeth Council offices; events that celebrated the site – these actions were all employed as acts of resistance to the Festival Wing plans.

This resistance to being moved encapsulates the underlying, perhaps most fundamental reasoning for the LLSB to resist the Festival Wing plans; indeed the slogan was 'You Can't Move History'. But there are a myriad of interlinked, composite and sometimes even conflicting lines of argumentation that further highlight why the 'Battle for the Undercroft' is microcosmic of the Creative City debate more broadly, and why it is such a critical process for the arguments put forward in this book. In the following pages, I want to outline why the fight for the preservation of the undercroft represents the current tensions that exist within the Creative City paradigm by exploring certain themes emanating from the battle between LLSB and the SBC, and the eventual 'victory' for the skaters. The historical importance of the site as a subcultural 'mecca' is clearly critical here, but the reasons why it has become so (i.e. through creative reappropriation) and what that represents in the wider urban context are critical to the search for a more creative city.

One of the issues with the whole LLSB campaign is that the Festival Wing plan is a clear case of a consumerism that is based upon a narrowly defined idea of culture. It is the latest attempt to create a Cultural Quarter; indeed the Festival Wing is to be part of what the promotional literature calls the 'South Bank and Bankside Cultural Quarter – the largest, liveliest cultural area in the world' (Southbank Centre 2013a: 1). As such, it is to become an identifiable area that is created for the purposes of cultural consumption. As we saw in Chapter 4, these Cultural Quarters have become commonplace across the world, and are the go-to format for cities looking to upgrade their amenities in ways that are seen as popular and easily packaged for entrepreneurial urban government systems. The South Bank in London is no doubt one of the country's most famous and commercially and economically successful cultural 'centres', but the expansion into the 'Festival Wing' and the corporate plans that it consists of are further reducing the broad cultural provisions on offer in the area. The reason for the development put forward by the SBC was the modernisation of existing spaces, but the main thrust of the narrative revolved around the 'arts and culture for all' slogan (seen in Figure 7.1). But when studying the plans, it is clear that the definition of the 'arts and culture' put forward runs along high-culture/mass-culture lines (a distinction described by Gans 1999). For example, the provision of extra spaces for orchestras, art galleries and theatre productions will be set alongside gig venues and dance clubs. Extra internal spaces will be given over to community groups and educational arts studios will be created to give 'opportunities for 45,000 school pupils each year, giving priority to those who have the least access to the arts' (Southbank Centre 2013: n.p.). The infrastructure that is proposed then is very much pegged to the production of a high culture-popular culture nexus (the positioning on which is obviously a matter of taste (Gans 1999)) because it is this nexus which aligns most with the political and financial goals. The urban development strategy that the SBC's plans are a part of requires a viable, pragmatic and instrumental product in order to maintain a level of control.

Chapter 4 detailed the ways in which contemporary urban development is predicated upon the commoditisation and subsequent control of urban space, and given that the provision of the arts and culture is now firmly part of such economistic determinism, the new Festival Wing plans are highly emblematic of this contemporary process of urbanisation. Such a process, however, necessities a homogenisation of cultural provisioning in order to lower costs and increase profit; it is fundamental Fordist economics in an urban realm. Lefebvre (2003 [1970]) famously outlined the 'industrial city', the politics of which are isolating, the power in which is concentrated and the aesthetics of which are increasingly homogenous and commoditised to facilitate marketisation. The Festival Wing plans represent the latter ubiquitously, a homogenised offering that replicates existing urban characteristics and reduces cultural offerings to tradable commodities. In essence, the Festival Wing plan epitomises the current agenda within Creative City building techniques.

Such an agenda also includes the marginalisation of certain activities that do not adhere to the particular view of culture put forward by the urbanisation narrative. Activities such as skateboarding are difficult to integrate into such a monetised conceptualisation of culture, not least because of its reactionary ethos and counter-cultural rhetoric (Borden 2001). However, if there is a way in which such activities can be capitalised upon, if they too can be 'produced' under the idiomatic renderings of the Creative City (i.e. homogenous replication), then they too will become part of such development plans. As such, the notion that the skaters (and the perceived 'associated' activities that have been rather patronisingly labelled as 'urban arts') could have been 'rehoused' in a designated area is part and parcel of the growing trend of cultural-led urban policy discourse to encourage the commercialisation of subcultural and subversive activity. Capitalistic urban development, which the Festival Wing represents, thrives on the predication of the unique and the novel, which subcultural and subversive activity represents. Being the first to commercialise a 'new' cultural activity, co-opting whatever is the latest iteration of 'cool' greatly increases the ability to command higher financial rewards. Put simply, if subversions can be co-opted then there is a good profit to be made. Boltanski and Chiapello (2005) have such an argument as part of their much theoretically broader and comprehensive critique of capitalism. They talk of the ability of capitalism to offer 'liberation' to the oppressed, but this only masks further degrees of control. They argue that:

> Capitalism attracts actors, who realize that they hitherto have been oppressed, by offering them a certain form of liberation that masks new types of oppression. It may be said that capitalism 'recuperates' the autonomy it extends, by implementing new modes of control. (Boltanski and Chiapello 2005: 425)

The SBC, with the proposed creation of an 'area for the urban arts', was attempting to do just that. By offering a new space, constructed under their own terms (financially and politically), they are offering a new alluring space, but one that would require the surrender of a certain power that the skateboarders currently have with the undercroft, gained through their decades of appropriation.

There was a comparable event in Philadelphia's Love Park which, like the undercroft, became a place where skaters would congregate given its 'perfect' skating architecture. As it become more popular, the city council passed a law prohibiting skating in the area. And despite the popular X Games being hosted in 2001 and 2002 in which skaters were allowed to skate, in the subsequent years, the city enacted more aggressive criminalisation strategies against them (Howell 2005). In 2003, Love Park was 'updated' and many of the 'perfect' surfaces for the skateboarders were replaced with grass, wood and rather more 'unskatable' surfaces. The marginalisation of skating from Love Park in a relatively short time stands in rather stark contrast to the longevity of the undercroft and, given the fact that the plans of the SBC actually included a 'new area for the urban arts', shows how subcultural activity has become more intertwined into commercialised cultural urban discourses (see later in this chapter). Indeed, the plans for the new area for urban arts under Hungerford Bridge (150m away from the undercroft) were extremely conspicuous in the Festival Wing promotional material. SBC has consulted with respected skaters and academics to devise an area that would have been conducive to skating and allowed the pastime to proliferate alongside the new developments. However, such a move highlights more ingrained processes of city control, economic determinism and political linguistics within the Creative City paradigm.

Also, the proposed relocation of the skaters to the skate park was an attempt to propagate a more tangible and 'identifiable' form of subcultural activity. By creating a skatepark from scratch, but one that is born from plans that are fundamentally aligned to London's attempts at global competitiveness, it would have given inexorable power to the instigators over those who use it. Also, rebranding it as an 'area for urban arts' is a rather patronising attempt to legitimise and forcibly collectivise a very disparate and tension-laden group of creative subcultures. The complicated historiography of the undercroft space, with its struggles over ownership, who has the right to use it and so on, are of course not free from questions of spatial power and politics. And as we have already discussed, it can be accused of normalising certain biases. But these are tied into the subculturalisation process more broadly. With the proposed creation of a brand new space that is within the framework of an overarching cultural place-making plan, any users would have sacrificed their 'right' to claim such a place as their own. Lefebvre (1996 [1968]: 158, original emphasis) noted that 'the *right to the city* cannot be conceived of as a simple visiting right or as a return to traditional cities. It can only be formulated as a transformed and renewed *right to urban life*.' As such, the *right* of the skater in this instance to actively remake the urban environment around them is completely eradicated. Purcell (2002, 2013) notes, when analysing Lefebvre's idea, that the 'right' which is being referred to includes more than simply the right to claim place in the face of an economic and/or cultural hegemony, but more the right to a continually open and transient city, where the constant search for a more democratic urbanity is fraught with tensions, contestations, rhythms and turbulence – what Lefebvre calls an 'urban society'. He argues that:

> Urban society is 'virtual' because it is not yet fully actualized. It is a possible society, one that is inchoate, emerging, in the process of *becoming*. But urban society is nevertheless real; it is operating right now, in the present.
> (Purcell 2013: 319, my emphasis)

The 'becoming' of urban society is theoretically distinct from 'being' and draws upon political and social theoretical postulating by Deleuze and Guattari (1987) articulated in the previous chapter. The very act of creating an environment that would have essentially been spoon-fed to the skaters takes away the ability of *becoming* and enforces a state of *being*. Such a place would no doubt have had tight security, enforceable management and laws, stricter control on what could have actually gone on there and an overall ability to change the 'rules of the game' at will. A skatepark in Manchester, for example (constructed under the Mancunian Way), that is owned and managed by a public-private partnership of Manchester City Council and ProjektsMCR is not only a purpose-built skatepark but now charges for membership (and recently doubled their membership fees (ProjektsMCR 2013)). Such an operation is a clear indication of the appropriation by capitalistic urban processes of creative subcultural activities. The new skatepark that was proposed at the South Bank then was an attempt to control the activity of skating within an apparatus that is conducive to profiteering and malleability. The promotional material by the SBC suggested that the cost of the new skatepark would have been £1m. There is of course little doubt that in order to recoup such a large outlay sponsorship of the space would have been offered, perhaps even naming rights. It therefore represents the SBC looking for stasis which is in contrast to the blurrier, more contested power relations that are inherent in the current undercroft skate spot. Yes, it has been there for over 40 years (in one form or another), but the uneasiness of the authorities, their attempts to stop them, the skaters' defiance of such practices, the subsequent tourist attraction, the incorporation of BMXers and inline skaters into the area and the reticence of the skateboarding community – all these issues and momentary events of (re)appropriation of unlegislated space are exemplars of the tumultuousness of becoming. The new area for urban arts would have eradicated these heteroglossic voices and presented a codified, tangible meaning to skateboarding – one that is profitable as a Debordian spectacle. For that is what a new purpose-built skatepark would have espoused – the reduction of skateboarding to a state of spectacle that can be determined economically. The current skate spot is already a place where passers-by will stop and watch. A stroll along the riverside promenade is not complete without the clatter of the skate wheels against concrete, and there are very few who will not stop, watch, take photos, even speak to the skaters at times. It is very much part of the aesthetics of the South Bank. However, in this case, such imagery and experience has been achieved through the emancipatory politics of the skaters themselves rather than the instrumentalism of the SBC and the Creative City. With the future of the undercroft secured, it remains to be seen whether or not any future plans will include such a park. The way that the Creative City policies operate, it would not be surprising to see future plans offer alternative skating arenas in an attempt to rival the undercroft. Again, of course, only time will tell.

The Creative City will no doubt lick its wounds, but certainly will return at some stage to reappropriate the undercroft and all the other localities of resistance in another more subtle, perhaps less direct way. It has to, as this is the nature of hegemonic creative urbanism. It constantly shifts its affectivities of PR, marketing and advertising to achieve the ultimate goal of profiteering. Creative subcultures can be resistive to it for a time, but the longer it resists, the louder it becomes, the more hierarchical it has to be, the greater the risk of appropriation. We have seen this with the increasing ossification of LLSB into a 'brand' to counter the SBC. Realising a creative city means constantly moving on, continually subverting functionalities, hierarchies and hegemony in search of the new. This is what it means to be creative. The placing of creative subcultures in particular locales, in some ways, defines this logic. Topophilia can take hold and begin to inculcate a wilfulness to maintain the existing subcultural order that overtakes desire-production. To paraphrase De Certeau, whatever we win, we should not keep. This is not a call to relinquish the undercroft, far from it. After all, 'you can't move history'. The things that we don't want to keep are those that mirror the Creative City and espouse their injustices, the systems of functionalities that urban subversive practices are attempting to eschew. The victory of LLSB against SBC was a triumph for community-orientated, grassroots subversive activism against hegemonic Creative City politics. The year and a half 'Battle for the Undercroft' encouraged people from very diverse backgrounds (in terms of age, race and class) to come together and form a commons – a body of political action. Continuing to champion sites of creative expression and communities of practice that are forming a contemporary urban commons that counters the prevailing urban hegemony is what we are encouraged to do in order to be democratic (Gibson-Graham 2006; Purcell 2013). And the fight for the undercroft was just that – it exemplified the fact that there are pockets of activism in the Creative City; it highlights that if the stakes are high enough, then places are worth fighting for. Moreover, its wider urban politics is a call to maintain urban subversion, to show fidelity to the creative truth that the undercroft engendered. It is a call to continually be truly creative at the undercroft and subvert systems, signs and hegemonies in all their forms.

As much as the undercroft allows us to see how a creative city can be made manifest, there are many other places embedded into the Creative City that have similar characteristics visually, politically and/or creatively. Some are, however, more easily mitigated against, while others are part of the Creative City from the start.

Vauxhall Walls

During my (all too brief) ethnographic study into parkour, there was a particular spot in London which has come to be a kind of 'hotspot' for practising parkour, namely the 'Vauxhall Walls' (Figure 7.4).

This area was originally designed as the communal part of the adjacent tower block, Haymans Point. It was built in 1966, at a time when the Le Corbusier-inspired 'streets in the sky' housing initiatives were being built all over London (and indeed the rest of the world). The 'garden' area that can be seen in Figure 7.4 was made up of low-level concrete walls, play areas for children and seating. Nearly half a century later, the concrete rectilinear design of the 1960s has become somewhat tired, out-of-fashion and neglected. However, the relics of these urban artefacts have inadvertently made for an almost perfect place to practise parkour (or its more English-language and media-friendly lexicon, freerunning). Parkour Generations, one of the foremost parkour institutions in the UK, regularly hold sessions there and it is often used in online parkour videos. It has, as many traceurs told me, become almost a right of passage for anyone who takes parkour seriously in London. The ideal layout is enticing for many people starting out, and provides a location where people are encouraged by other users (known or not) and has a sociality that not only encourages

Figure 7.4 Vauxhall Walls parkour spot, South London.
Source: Author's photo, 19 February 2013.

'beginners', but one that is fundamental to the proliferation of parkour as a subculture. Like the undercroft, the space has a tacit code of ethics that are adhered to (more or less), and while it does not have the longevity of the under-croft, through word of mouth, online dissemination via community forums, videos and user-generated maps it has become one of the more 'famous' of parkour spots in London.

Much like the undercroft, it is also contested space. Haymans Point is a resi-dential area maintained by the Vauxhall Gardens Estate Residents and Tenants Association (VGERTA) along with Lambeth Council, and over the years the traceurs (or freerunners as they are referred to in the majority of the council's and residents' literature) have been largely unwelcome. Signs have been put up insist-ing they stop (see Figure 7.5) and I have been told that often residents will shout from their windows or when passing by, insisting that they desist (although I never experienced this directly).

Derogatory insults are apparently the norm, and despite the traceurs attempt-ing to extol the virtues of parkour, they are threatened with police action (even though what they are doing is not illegal) and one participant told me that he was even threatened with physical violence. Overall though, the traceurs are viewed as 'anti-social'. Moreover, VGERTA refer to the Vauxhall Walls area as the 'sunken pit', and in August 2014, secured funding for 'redevelopment' of

Figure 7.5 Sign marginalising freerunners at the Vauxhall Walls.
Source: Author's photo, 5 August 2014.

the site into landscaped gardens. A post on the SE11 Action Team blog (from 2013) notes:

> Coverley Point and Haymans Point are two 1960s towers on Vauxhall Walk, which are part of the Vauxhall Gardens Estate. They are designed with large sunken concrete areas around the base of the towers. 'The pit' outside Coverley Point has suffered badly from antisocial behaviour – particularly from large groups of 'freerunners' [or traceurs] congregating and disturbing residents. Lambeth Living and representatives of VGERTA are keen to pursue a scheme to 'green' the sunken area facing Vauxhall Walk, reducing antisocial behaviour and greatly improving the attractiveness of the local environment.

This wish has now been granted and the site is now being redeveloped as a more 'attractive' garden that the residents will be able to enjoy but the traceurs now will not. The language being used by the council, local residents associations and property developers is very much aligned with that which was discussed in Chapter 3, in that the area is need of '*re*development' or re*vital*isation. Such phrasing is symptomatic of the way in which urban planners, property developers and financiers achieve their goals of profiteering (Lees 2003b). Here, we see the Vauxhall Walls area been designated as a sunken 'pit', engendering distaste and a sense of unworthiness. Clearly if an area is filthy enough to be considered a 'pit' then it is in need of beautifying. There were voices from the local residents who often complained of the noise and anti-social behaviour of the traceurs and there was a consultation process (unlike many of the larger developments, such as that seen in Liverpool Waters discussed in Chapter 3) that took such views into account. However, from the point of view of the traceurs who used the Vauxhall Walls (and from my own experiences of using it), the area is anything but a 'pit' instead, it is a place of meticulous preparation, a site of triumph and achievement, and a place that where personal histories were made. As a subculture, parkour was appropriating this 'dead space' (in the sense that the residents were not using it given its concrete aesthetics had fallen out of fashion) into a site with alternative functions. This reappropriation can be seen with the following outtake from an ethnographic study of parkour at the Vauxhall Walls in 2009:

> I recall a night-time training session in Vauxhall, South London, in 2009. Meeting outside the Royal Vauxhall Tavern, around ten traceurs of differing experience took off running, like a wolf pack, following our leader. Eventually we came to a cluster of tower blocks built in the affectionately known 'brutalist' style. We began training around the courtyard [the pit], which consisted of a number of high walls that determined the movement of the residents in a seemingly arbitrary way. At a certain point I remember being asked to perform a 'cat leap', that is, a running jump from a narrow ledge, to land hanging from another ledge. The drop must have been over twenty feet. While nervous, I eventually executed the movement. I became

aware, in that moment, of the oppressiveness of the architectural design in this council housing. It is only in using a wall in a paradoxical way that I understood its *function* – only by encountering the wall as a danger, a twenty-foot drop, was I able to question: Why is this wall so high? Who is it trying to keep out? Or in? (Chow 2010: 151, my emphasis)

Such 'paradoxical' usage of the wall was desire-production realising an alternative function, and hence began the questioning of the existing functionality of the wall as a containment device. Exploring different functions of the Vauxhall Walls is a key facet of parkour's engagement with that particular place. The continued usage of that wall as an obstacle to do cat-leaps on inculcates a functionality of its own, the subculturalisation process in action. This also engenders the beginnings of topophilia. The traceurs I practised and trained with at the Vauxhall Walls began to take care of the place, removing rubbish so that we could train more safely, cleaning up after training and maintaining the site as a safe place to practise. We were realising a more creative city by utilising the topology of the site to satisfy emotional and affective desires. By desire-producing, we were creating new functionalities of signs based around the spectacle of parkour; the virtual circulation of the imagery and interaction with the location helped to serve the ossification of this site as an important location to the subculture.

But now the Vauxhall Walls are being consigned to history. The traceurs that currently use it will move on and the once subversive place will be subsumed into the capitalist urban realm. There is no petition to save it, no viral campaign that garners international attention, no mayoral intervention. It seems that you can indeed move history, as long as it is a short enough history. However, parkour, like skateboarding, has begun to build purpose-built parks. London has the LEAP park situated near Paddington station, which is a collaborative project between Parkour Generations, the City of Westminster and Parkour UK. It is a place where people can pay to train safely with professional guidance. It has spawned a proliferation of parkour 'parks', some of which are free to use as a service provided by the city (just like children's playgrounds), others which are more intensively designed and maintained, and are part of a business venture. Either way, the *placing* of parkour has very much become engrained in the creative subcultural provisioning of cities. I have argued elsewhere that this kind of 'institutionalised' parkour belies the original philosophy of the subculture, because a core truth of parkour is in its capacity to defy prescribed usages of the city (Mould 2015). The traceur has often been likened to Benjamin's flâneur (and espouses very similar accusations of gender and able-bodied bias (Kidder 2013)). As Atkinson (2009: 174) notes: 'The late modern urban traceur movement is an assemblage of flâneurs who bound across and scale city spaces to underline the oppressive nature of them.' Also, I have argued elsewhere that traceurs are smoothing out the striated space of the city, creating lines of flight from the urban apparatus of capture (Mould 2009). It is an intoxicating vision, the traceur as a free spirit, breaking from restrictive urban architecture, jumping across rooftops and forging a different and individualised path through the city. There is no doubt that parkour engenders this philosophy, its original name, *l'art du placement*, anglicised as the 'art of displacement', is highly literal – traceurs are displacing themselves from the city via a corporeal artistry that is both spectacular and emancipatory. Yet, there are intensive debates and counter-arguments from within the parkour community and from observers and researchers, all with a view on where parkour 'should' be practised. Some have noted the counter-intuitiveness of parkour parks (Ameel and Tani 2012; Gilchrist and Wheaton 2011), others have argued that while this is a core ethos, as parkour gains international popularity and more people practise it, they all need safe places to train (see Edwardes 2009). This debate is again an example of the kind of malleability that comes about through the subculturalisation process. As parkour begins to become more recognised, as it is integrated more into the capitalistic spectacle by being more visible online, on TV and in Hollywood blockbusters, its placing will become more 'institutionalised'. The growth of the Creative City paradigm comes about through the institutionalising of these subcultures. The capitalistic mechanisms that are in constant need of reinvention thrive on the appropriation of subcultural activity.

The 'freerunning' that will soon cease to be conducted at the Vauxhall Walls is being marginalised, and its history as a parkour training spot will soon be complete. The 'pit' will not have the opportunity to continue to be woven into the subculturalisation process of parkour; it will forever be remembered rather than lived (assuming the development into landscaped gardens goes ahead as planned). One suspects that had the place been given more time to germinate, to ossify history and concretise its *place* in the subculture, then its redevelopment may have 'fought' with the same rigour as the undercroft. There are of course large differences, the main one of which is that this is a residential area. The local residents can claim not only legal status of the land, but also I noticed a much more subservient attitude with the traceurs as compared to the skateboarders. Perhaps this can be put down to the more 'serene' philosophies of parkour. It has strong relational ontologies with spirituality and East Asian contexts. Many traceurs have trained in martial arts and the early videos contained overt 'orientalist' symbology (Mould 2009). More readily, however, this is primarily a residential area. It is not claiming to be a place that provides 'culture for all' and there are unwritten 'rules' of subservience to local residents. Theorising place as assemblage (Creswell 2013) then the Vauxhall Walls is made up of processes and agencies that are primarily residential. The laws governing it, the architecture, the behaviour *in situ*, the tacit understanding, the sensorial qualities – they are all engendering a 'residential atmosphere'. It is easy to sympathise with the residents as to why they want to marginalise the traceurs. There are multiple forces at work which can help (or hinder) the marginalisation of particular creative practices and therefore the creation of an identifiable subculture. Had the traceurs reappropriated a more 'publicly accessible' place then the residential qualities of the place would have been less distinct. If the perfect parkour architecture of the pit had been found in a 'dead space' (like the undercroft) then a different story may have emerged. The tensions between the traceurs and the residents added another layer of narrative to the pit, one that will further sediment the palimpsest of that place, whatever it ends up being called.

Parkour does not have as long a history as skateboarding. The Vauxhall Walls are more residential than the undercroft. The traceurs do not feel as much symbolic attachment to that place. There are a number of reasons that can be teased out as to why the Vauxhall Walls have not 'succeeded' in stopping their redevelopment. But to compare them in this way is unhelpful, arbitrary and part of the processes of hegemonic control. 'Success' is a concept that only exists within the Creative City. Instead, a creative city sees them as one and the same, through their creative and subversive functionality. The alternative functions that have been realised at the undercroft and at the Vauxhall Walls means that they are forever to be connected as part of the creative city. They are places that have had their capitalistic functionalities subverted; the temporal longevity of that subversion is unimportant as it catalyses the subculturalisation process and the inequalities that it brings about. We should instead be thinking about these places as moments of subversive creativity, situations that have brought about experience of the city beyond the spectacle, beyond that which the Creative City has prescribed us to. It is this situationism that is to be celebrated, it is this that is a creative city.

But what of places that try to manicure such instances of subversion? What about those places that are set up to invite subcultural activity? Can we think of these places as part of the creative city too? I want to use the example of Leake Street to ask that very question.

Leake Street – graffiti tunnel

About a mile away from the Vauxhall Walls and a mere half a mile from the undercroft is another place of urban subversion. Leake Street, or the graffiti tunnel as it has become known, is one of the UK's first 'legal graffiti sites' (see Figure 7.6). The oxymoronic nature of this term is yet another microcosm of how the Creative City can begin to reappropriate the very activities that were once 'outside' it. Unlike the undercroft though, Leake Street was not appropriated by street artists in opposition to the owners. It doesn't have the same history of contestation and as such does not invoke the same topophilia. Moreover, graffiti tunnel can be seen as a physical and *placed* manifestation of the process of the Creative City reappropriating subcultural activity. This is because graffiti tunnel came about through a direct response to the famous street artist Banksy's increased popularity and co-option into the commercialised art world (as discussed in the previous chapter).

Fresh from the 'Barely Legal' show that he put on in Los Angeles, Banksy set about creating a new show in London, but he needed an appropriate space. In 1994 when the Eurostar train terminal was built at Waterloo Station, Leake Street (which ran under the platforms) was the taxi rank. But since the Eurostar moved to St Pancreas station in 2007, the taxis have moved on and very quickly the site became a place where homeless people would gather for shelter, graffiti would be

Figure 7.6 Leake Street or the graffiti tunnel.
Source: Author's photo, 19 February 2013.

conducted, and it generally became a place of marginalised legality. For Banksy, it was the perfect place. In May 2008, the 'Cans Festival' opened for a period of six months after being 'granted permission' from Eurostar (who still owned the site). During this time, Banksy exhibited some of his latest work as well as inviting street artists from all over the world to showcase their art. At this time, street artists were still operating covertly as there were minimal 'outdoor' streetscapes where they could practise without fear of being caught. They could operate indoors in private, but conducting street art in public spaces was still illegal. Leake Street therefore was one of the first 'legal' graffiti sites in London (possibly the world, although given the complexities of defining the legality of these sites it is difficult to ascertain the official 'start' of the legalisation of graffiti in any specific place). It allowed people a place to come and practise street art without any need for clandestine activities at all. There was no need to be subversive.

Leake Street as a place of 'legal graffiti' has been crafted for the purposes of the proliferation of street art and its co-option into the Creative City. Six years on from Banksy's Cans Festival, it is still a site of legal graffiti and is frequented daily by people spray-painting on the walls. Those who own and manage the space (Eurostar, then Network Rail, the UK's governing company for the rail infrastructure network), the council and the police are all complicit in the activities of Leake Street. There is even a 'code of conduct' that is marked high up very clearly upon entering (see Figure 7.6). 'The Tunnel, Authorised Graffiti Area' is very prominent before entering Leake Street, with the rules of 'No Sexism. No Racism. No Adverts' made clear. It is hence clearly demarcated with explicit rules, codes, guidelines and ethics. I have witnessed times of conflict, with some street artists arguing over which spaces they are painting on, and whether or not the messages are offensive or not. Some of them threatened to spill over into violence. This, thankfully, never happened given that the majority of artists would intervene and the situations would resolve themselves democratically. This kind of community-level management is to be celebrated as a viable way of conducting creative subcultural activities. However, the space is also policed somewhat heavily. Police cars would be present in the majority of my visits with homeless people moved on and larger groups of people who were 'loitering' often questioned at length. While Leake Street encourages street art and 'legalised' graffiti, it still attracts marginal activity (homeless people, drug-taking, public drinking, 'anti-social behaviour' and so on), although even this is now beginning to change.

Six years on since Banksy's Cans Festival, the area has attracted a number of more commericalised 'traditional' kinds of creative activity. Rather tellingly, a press screening for *Exit Through the Gift Shop* (2010) was held in one of the indoor spaces off Leake Street. Also, there is an exhibition space called 'The Vaults', also just off Leake Street. Until 2012, it was a storage space for Network Rail, but after a successful 'one-off' event by Punchdrunk (an 'immersive' London-based theatre company that has conducted innovative and participatory shows in the UK and the USA (Machon 2013)), the space has been turned into a permanent exhibition gallery. It now holds regular 'pop-up' events such as concerts, film screenings, plays and art exhibitions that are immersive, surreal, participatory and 'non-mainstream'. Notwithstanding the now pervasive use of the term 'pop-up', the Vaults, along with the subversive but 'legalised' characteristics of Leake Street, are beginning to formulate more permanent, commercialised Creative City activity that is inherently place-based. It is, in effect, becoming a (sub)Cultural Quarter. It started out as a space for Banksy's exhibition which would not have come about without his co-option into the commercialised and formalised Creative City framework. The identification of Leake Street as a specific *place* in which to do street art signalled the beginning of the subculturalisation process of Leake Street, which is moving it toward being a place of commercialised subcultural activity. Further evidence of this can be seen in the recent completion of a skatepark in the area immediately south of Leake Street. Called the 'House of Vans', after Vans the highly successful US skateboarding apparel company, the site opened in August 2014 and features a purpose-built skatepark, an exhibition space, regular music shows and other creative cultural activity. The planning application for the site, which is available online (Lambeth Council 2014), includes the various safeguards that had to be put in place in order for the site to be established. Mitigation against flooding, how the bins will be taken out, access, security and other measures all needed to be outlined. Section 8.11, which details the consultation with the Designing Out Crime officer, states that:

> The activation of these frontages will be a welcome introduction of activity and surveillance in this area. The arches are the Property of Network Rail and therefore the structure and Leake Street comes under the remit of British Transport Police. It is desirable to improve activity and surveillance in this area which could in turn reduce vulnerability to crime and anti-social behaviour. A robust management and maintenance plan will be necessary to mitigate against local crime trends, inappropriate behaviour, nuisance, litter, damage & graffiti.

The frontages refer to the access to the site, which is via Leake Street. The opening of the skatepark and exhibition space then is being welcomed because it 'mitigates against' local deviant behaviour (including, interestingly enough, graffiti) and provides further instruments of surveillance, management and control. Its doors opened officially in August 2014, and on some preview visits to the site days before its opening I saw that there were in place overt CCTV cameras, obvious directional signage and wall-high advertising and branding. In addition, it is interesting that this space represents all the same characteristics of 'urban arts' that were being espoused by the SBC at Hungerford Bridge. This makes the creation of the 'area for the urban arts' by the SBC as a ploy to simply displace the skaters (rather than any actual desire to benefit the skating community) from the undercroft even more apparent.

The area around Leake Street then is becoming part of the Creative City as an identifiable place of creative subcultural activity. It is formulating identifiable geographical, legal, cultural, artistic and coherent boundaries. An assemblage of institutions and individuals including the local council, the police, private companies, invested individuals, complicit artists, Network Rail, the City of London and others come together within frameworks of planning, procedure, finance, real estate (and so on and so on) to create areas that are directly part of the commercialising, profit-making, capitalistic Creative City. Through the fluid negotiations, the contested relationships and the jostling for power, these various institutions and actors are formulating new concretised modes and spaces of capitalistic accumulation. Neoliberal practices, as Chapter 2 discussed, are far from static and linear. They are better thought of as the manifestations of particular assemblages that react to the conditions and specificities of place (Springer 2010; Davies 2014). Ong (2006: 14) notes that as 'an array of techniques centered on the optimisation of life, neoliberalism migrates from site to site, interacting with various assemblages.' Therefore the processes that are currently going on in Leake Street, the institutions that are formalising and commercialising once subversive creative subcultural activity, are the temporarily coagulated forces of capitalistic change. They come together to create new physical spaces of commercialised activity because, as has been discussed throughout this book, it is through the claiming of space that the most profit is made. Leake Street therefore is in the act of *becoming* a Cultural Quarter. Yes, it is housing activities that have their foundations in illegal, resistive and counter-cultural activity, but it is precisely these activities that are attractive to the Creative City because they espouse 'newness'. The subversiveness of these activities have been largely eradicated. Like the gentrification of Stokes Croft in Bristol, Leake Street is inculcating a cool bohemian, edgy aesthetic that is seen as attractive to the creative class, and so it is no wonder that the larger Waterloo Master Plan aims to continue the development of this site with the usual luxury flats, business units and upscale leisure facilities. Unlike the undercroft that is fighting for the non-commericalisation of the space, Leake Street, while espousing a similar subversive aesthetic, has no *true* creativity to fight for. I have argued above that a creative city is one that celebrates places like the undercroft because it represents the continued flight from the apparatuses of capture that attempt to commercialise it, maintain subcultural hierarchies and biases, and divert its subversive origins for structural and financial gains. In contrast to this, Leake Street, precisely because it was 'born' from the highly subculturalised *being* of street art, cannot be part of the creative city. Through no fault of any one individual (although Banksy could arguably have been the 'tipping point'), what Leake Street is, is exactly the kind of place that the Creative City craves.

Summary

This chapter has highlighted three specific places of the 'sliver of subversive spaces' in the South Bank of London. There are many other places that could have been identified and brought into the discussion in slightly nuanced ways. Spinney (2010) has written about the BMX and trial-riders of the South Bank, predominantly operating in the Shell Centre. Like the Vauxhall Walls, the architecture of the Shell Centre, Spinney (2010: 2922) notes, is 'exactly the kind of area that BMX, trials riders, and skaters would find attractive'. The 'loose, pseudo-public space' of the Shell Centre means that they trial-riders are often approached by security guards or members of the public and told to desist. But far from embodying an ethos of resistance, Spinney found that this was largely absent from their practices. He concludes that 'that practices can be inappropriate but still *appropriate* spaces without rejecting or resisting dominant uses' (ibid.: 2933). Within the subculture of BMX and trial-riding then, the South Bank is a particularly important space. Yet there the performance of the riders lent itself to the overall spectacularised aesthetic of the Southbank Centre. Also, a bit further down the river, we have Battersea Power Station. It is an iconic structure in popular culture, popularised in films, television shows and books and is the subject of many contemporary psychogeographers. According to many urban exploration accounts, it is relatively easy to access (i.e. it has limited surveillance and the boundaries are porous) and has been the subject of many online photo essays. Garrett (2013) describes it as his 'dark princess'. It is seen as a 'training ground' of infiltration, somewhere that explorers can go to get a taste for the activity. Given the relentless march of London's urban governance to develop the city for capitalist gains, even this iconic structure has been earmarked for conversion into luxury flats, shops, hotels, leisure facilities, etc. Therefore it can also be viewed as yet another space that has been appropriated by subculturalisation processes, given further histories, stories and narratives, but then finally succumbing to the powers of capitalistic development.

Another place worth briefly mentioning is 5pointz in Long Island, New York (see Figure 7.7). 5pointz was another legal graffiti site that appropriated a disused factory somewhere around 2002 (Kramer 2010). From that point, the walls both inside and out were covered in colourful, ornate and highly artistic street art murals. It was a tourist destination and became an iconic site for street art in a city that has a very protracted history with graffiti. Like Leake Street, people could come and practise their art without fear of recrimination by the police.

Figure 7.7 5pointz 'legal' graffiti spot, Long Island, New York City.
Source: Author's photo, 17 March 2013.

However, as of August 2014, the site was whitewashed and then torn down, to be replaced by luxury flats in the ever-increasing hyper-gentrification of New York. Petitions were set up to save it, but unlike with the undercroft, there was little people could do in the face of belligerent private landowners. What this represents, however, is the continual movement of places of subversion. By continuing to be in place, it only risks further ossification which can engender the normalising biases of street art. Tearing it down forces the street artists to re-engage with the initial desire-production that saw them start the practice in the first place. Yes, New York loses a beautiful site that inspires people to think differently about their city; it will lose a site where people can practise street art (relatively) safely; it will lose a site of iconic global subversion. However, it does mean that new places will be sought, new functions will be realised and new urban subversion will occur.

A similar story has occurred in Shanghai. M50, the creative zone detailed in Chapter 4, is actually a nod to its address, number 50 on Moganshan Road. The road has other cultural institutions on it (art galleries, exhibition spaces, etc.) but it is also the city's first (and only) legal graffiti site (Figure 7.8). In a city where graffiti is criminalised heavily, it is unsurprising that this graffiti wall has also been earmarked for demolition and in its place the construction of more

Figure 7.8 Moganshan Road, Shanghai, or graffiti alley.
Source: Author's photo, 14 October 2012.

restaurants, cafes and retail units. By the end of 2013, half the wall had already gone (Minji 2013). The artists are largely seen as inconsequential to the rampant development of Shanghai, and one suspects that any campaign to maintain the site would be quashed rather quickly. Indeed, one of the developers of the site has been quoted as saying 'these graffiti painters just paint for fun, so pulling down a wall is no big deal' (Yin 2011: n.p., quoted in *Time Out Shanghai* 2011). Whether or not they 'paint for fun' is questionable given some of the political sloganising seen on the walls during my visit. Also, such an attitude is a rather brazen degradation of a skilled art form. Like 5pointz, Moganshan Road became a focal point for street artists in an increasingly restrictive city, but now they will be forced to continue their craft in other places, subverting further functionalities in the process.

London, New York, Shanghai – these cities and many more have rich selections of subcultural spaces, each with their own stories, histories and relationships to the Creative City. Empirically investigating them teases out the specific stories and histories of these places and the urban subversion that went on to manifest these spaces. However, they are fleeting, temporary and 'of a time'. In terms of the undercroft, the Festival Wing plans that included its demolition were released on 7 March 2013, but by 18 September 2014, the undercroft was saved. So in an 18-month period, the Battle for the Undercroft had been fought and a small victory won (although the war, I suspect, is far from over). Furthermore, the fact that the Vauxhall Walls redevelopment plan was announced and given the green light and the 'House of Vans' skatepark was planned, built and opened during the process of writing this book shows how empirical situations that speak to the constant co-option, appropriation and/or resistance of this can become out of date relatively quickly, such are the processes we are dealing with. But whatever the sites, whoever is being researched, however they are recorded, they all speak to the insatiability of the 'spirit' of capitalism, and the rapidity of the Creative City's appropriating mechanisms. More importantly though, each of these examples (and the many more than could have been used and will become apparent in the future) tell us about how a creative city can be realised, and yet how it can also be co-opted. With Leake Street in London, we are even witnessing the creation of a subversive 'zone', a (sub)cultural quarter in waiting. Such ossification of subversive activity only serves to neutralise its creative potential because, as has been evident throughout this chapter, the creative city has an uneasy relationship with the physicality of the city. De Certeau (1984: xix) tells us that tactics do not claim spatial ownership; instead, 'the place of the tactic belongs to the other.' Urban subversion allows the taking of places momentarily, but as soon as they begin to claim spaces as their own, they become susceptible to the strategic mechanisms of the Creative City. Staking a *claim* for a place creates forms that can be identified, it creates politics that can be contested, it creates functionalities that adhere to a system of signs, it creates codes of ethics that can be misdirected, it creates guidelines that can be misinterpreted, it creates biases that can inculcate inequalities, it creates subcultures that can enmesh with the processes of the Creative City and the injustices it catalyses. But, as was highlighted by the case of the undercroft, if we continue to realise a creative city and engage in urban subversion, then these potential risks can be mitigated. We need always to look to subvert those marginalising systems that are being formulated. We must engage in lines of flight and seek out new connections, new shared experiences, new moments of sociality that enliven the creativity of places. We need to continually become-minor, search for the margins and minorities that push away from centralising forces. We must continue to seek out new possible functions of objects. We must continually be desire-producing.

8   Creative City trajectories

A working title for this book was *Creative Cities: What They Are and How Not to Be One*. Such a title was concocted out of frustration with the lack of creativity, the social injustices, the economic determinism and the intense political mobility and homogeneity of the Creative City paradigm. In nearly a decade of working in and researching the neoliberal creativity agenda, it has become abundantly clear that the Creative City paradigm does not stimulate *true* creativity among people. It works upon the continued replication of a singular, narrowly defined view of creativity, one that creates financial profit for the few. And it has spread across the globe with consummate ease. Pushed by consultants, think tanks, politicians, academics and business leaders who stood to gain financially from its proliferation, the Creative City paradigm (and all the other creativity-inspired related ideologies) is an off-the-shelf, go-to policy for cities (of all sizes and locations) that promises vast profits. We saw this in Chapter 2, with the Creative City detailed as the next logical iteration of urban development because it paradigmatically brought together related and enmeshed ideologies of neoliberalism, the Global City, urban branding and the creative industries. Chapter 3 highlighted how the creative class and its incessantly pervasive political mobility catalysed a monumental shift in the way urban governments went about justifying existing development agendas. Then Chapter 4 detailed the way in which the mobilisation of the Creative City paradigm physically created zoned 'quarters' of cultural consumption and creative economic activity, a necessary step to realise the vast profits that the Creative City promises. Therefore Creative City policies such as rebranding, focusing on creative industry activity, attracting the creative class, building Cultural Quarters and Media Cities, beautifying public spaces around cultural and creative rhetoric, generally espousing an aura of 'cool' – they have all become all too obvious markers of a homogenous strategy of capitalistic accumulation. The evidence of their success at generating innovation and economic growth has been partial and contested, yet their gentrifying tendencies have been well noted. Yes, the real estate developers have profited from the construction of further business, leisure and residential facilities, but the increasing securitisation and marginalisation of 'other' kinds of creative activity from such spaces actually reduces the richness of creativity from cities rather than fuels it. Cultural participation among elites has no doubt increased for very specific kinds of consumption-orientated culture, but cultural desertification for the majority has been the result. There was a need therefore not only to highlight the problems of the Creative City (something which has been achieved highly articulately throughout the academic and policy literature of late), but to actively encourage alternative ways of thinking creatively in the city. Moreover, there was a need to realise that creativity is stimulated, but in reaction to the Creative City via urban subversion rather than as a part of it. As such, there was a need to counter the instrumental 'guides' of the Creative City with alternative descriptions, articulations and manifestations of creativity in urban spaces. There was the need to highlight the fact that a creative city can exist, but only if the Creative City paradigm is not only critiqued and dismantled, but a *viable* alternative is offered (hence the rather instrumental, tongue-in-cheek working title).

But what exactly is the alternative? If we are 'not to be' a Creative City, then what exactly can we be? How can we continue to be creative and not fall victim to the appropriating and co-opting forces of the Creative City? How can we continue to create meaningful new forms of knowledge, subjectivities, experiences and encounters without falling into a hegemonic trap? Can we realise a creative city? Chapter 5 began this quest by outlining a number of foundational theories, vernaculars, philosophies, thought-experiments and ideologies that are important tools in the exploration of a creative city. Chapter 6 then detailed how urban subversion as a process can help, but also has the potential to hinder the flight from the Creative City's apparatuses of capture. In Chapter 7, London's South Bank sliver of subversive spaces were exemplified as ways in which urban subversion creates different kinds of 'subversive' places, some of which contribute to the development of the Creative City just as much as they resist it.

Therefore what this book has shown is that a realisation of a creative city is possible, but is fraught with difficulty and risk. By resisting the Creative City, by critiquing the hegemonic processes, the risk is of ossification into an identifiable 'form' (i.e. a subculture) that is rife for the picking by the capitalistic apparatus of capture. Boltanski and Chiapello's (2005) monumental work charts how capitalism has become highly adept at absorbing such critiques (both social and artistic). Indeed, they argue that:

> Because critique makes it possible for capitalism to equip itself with a spirit which, as we have seen, is required for people to engage in the profit-making process, it indirectly serves capitalism and is one of the instruments of its ability to endure. This poses some serious problems for critique, since it easily finds itself faced with the alternative of being either ignored (and hence useless) or *recuperated*. (Boltanski and Chiapello 2005: 490, my emphasis)

The 'spirit' of capitalism has over the course of two centuries motivated people to continually serve its accumulation. Such accumulation though, they argue, is dependent upon the continual recreation of subjectivity. There is a continual replicative capacity (spirit) of capitalism that transcends humanity's capacity to produce surplus value. Capitalism, they argue, does not simply 'force' people to work – its spirit is subtle, it coerces, it creates conditions that allow critique, it allows the preservation of a critical distance (family life, civic societies, charity work and so on). By allowing such 'distance', it creates systems of its own accumulation, it creates deficiencies which can only be satiated through further profit-making processes. Therefore, according to Boltanski and Chiapello (2005), critique is an essential part of capitalism's accumulation process and therefore systematically creates it. Mouffe (2013: 73) has argued:

> To be sure, Boltanski and Chiapello never use this vocabulary, but their analysis is a clear example of what Gramsci called 'hegemony through neutralization' or 'passive revolution', a situation where demands which challenge the hegemonic order are appropriated by the existing system so as to satisfy them in a way that neutralizes their subversive potential.

Here, Mouffe is building upon the ideology of Boltanski and Chiapello and articulating how the subversive potential of challenges and critique to hegemony is neutralised and whitewashed. Boltanski and Chiapello also suggest that, over the years, this process has been further refined, with the post-May 1968 'spirit' of capitalism having the most efficient mechanisms at absorbing critique.

This book has detailed how the Creative City has refined this process even further. Part I detailed how the language of creativity has acted like a veneer to neoliberal capitalism that makes the recuperation of critique all the more palatable (Peck 2005). In other words, by labelling a city as 'Creative', it has effectively disabled and disarmed artistic and social critiques, and engineered a system whereby critiquing hegemony risks feeding into it. Boltanski and Chiapello (2005) noted critique is either ignored or recuperated, and Part II of this book argued that through the process of subculturalisation, these happen sequentially an/or in tandem. By othering and marginalising creative activity, the Creative City creates the possibility of recuperation by allowing such creative activity to proliferate and form a subculture. And now that edgy, bohemian, cool and counter-cultural aesthetics are actively encouraged as part of the Creative City, it is easy to see why such identifiable subcultures are amenable to recuperation. This constant 'search for the new' is part of the neoliberal strategy of urban entrepreneurialism (Harvey 1989) and is the driving force of the Creative City. Moreover, it catalyses the instrumentalisation of creativity into a packaged, politically mobile paradigm which has inherent injustices and inequalities built in. It is reducing critique (urban subversion, resistance and transgression) to mere amendments of capitalistic accumulation. This therefore is the summation of what I identified in the Introduction as the original *problem* of the Creative City.

To disrupt this process, therefore, this book has argued that there is the need to engage in urban subversion. This is the process not only of subverting the dominant ideologies of the Creative City, but it is also the process of questioning the subsequent hegemonies that such subversion can create. Engaging in urban subversion is to constantly 'become-minor' (Deleuze and Guattari 1987) to seek out the margins and the edges of hegemony, to probe the weak points and flee the apparatus of capture. It is to realise a creative city. As Part II detailed, such a process can have vastly different outcomes and create very different kinds of 'subversive' subjectivities and places. Urban subversion can create subcultures that replicate gendered, ablist and class biases, yet it can also create progressive urban politics that ignites activism. It can create places that stand for resistance and counter-culture, but also places that represent the 'new' latest iteration of the Creative City. It can be moments of insurgency that espouse substantial critique of the system, but also enhance neoliberal tendencies via advertising, branding and becoming spectacle. Throughout this book I have argued that there is the need to realise a creative city, to continually flee from the apparatuses of capture. But what will such a creative city actually look like? Chapter 7 explained how placing subversive activity results in different kinds of spaces, but we can extrapolate the realisation of a creative city to a more theoretical plane, because there is a multitude of ways in which urban subversion can manifest itself in a creative city, each with their own problems and potentialities. Some are of course more equitable and desirable than others. In the remainder of this concluding chapter, I want to posit a number of ideological and theoretical urban trajectories or 'scenarios' that could be obtained through practising urban subversion to a more or lesser extent. Each scenario is essentially a thought-experiment, a description of an utopian/dystopian/heterotopian idyll. They are not predictions, nor are they manifestos of urban futures. They are ontological descriptions of cities that would theoretically manifest through the different intensities, rhythms and subjectivities of urban subversion.

The Creative industrial city

The 'industrial city' was, according to Lefebvre (2003 [1970]: 35) 'a phantom, a shadow of reality, a spectral analysis of dispersed and external elements that have been reunited through constraint'. He was unrelenting in his assertion that the creation of urban space is becoming (and now, arguably, has become) the driving force of the mechanisms of capitalist control and production. He states that:

> The reality of urbanism [is that it] modifies the relations of production without being sufficient to transform them. Urbanism becomes a force in production, rather like science. Space and the political organization of space express social relationships but also reacts back upon them. (Lefebvre, 2003 [1970]: 25)

In other words, Lefebvre, in conceptualising the transformation of the urban condition to the 'industrial city', has inextricably linked Marx's ethos of production to urbanisation, producing a Fordist ideal of the manufacturing of cities. In effect then, what we have are cities falling off the end of the production line, or to use a more contemporary metaphor, being bought off the shelf, flat-packed and ready to go, much like the Cultural Quarter policy that was discussed in Chapter 4. Coupled with his exploration of 'abstract space' which is '*not* homogeneous; it simply *has* homogeneity as its goal' (Lefebvre, 1991: 287, original emphasis), these industrial cities of the contemporary era are 'abstract' in that they are unreflexive to local social specificities, cultures and politics, and aim to homogenise the urbanisation process through a fundamental capitalist ethos: low unit cost of production yielding greater profits.

The Creative City strives for such homogeneity. Its capitalistic mechanisms are designed to seek out the new (as Boltanski and Chiapello (2005) have noted), and those practising urban subversion offer such novelty. As has been discussed in Chapter 6, these activities are initially marginalised from the city, and hence proliferate 'outside' capitalistic development. In the creative industrial city, however, such marginalisation does not take place and the act of subversion is brought into the narrative of the Creative City almost instantaneously. The act of marginalisation is costly. Defensible architecture is expensive to implement and enforce, security patrols are wasted labour, securitisation of private land is utilising resources that could be producing more surplus value. In this city, those who engage in desire-production and subvert existing and established functionalities, those who realise an alternative function of the urban terrain are instantly brought into the capitalist mantra. They are offered contracts, given a wage, branded and hawked as new subjectivities that maintain the city's competitive edge. The smaller the 'time lag' between the realisation of something new, an alternative function, and the ability to profit from it, the less time resources are spent on guarding against it. In this way, it is a more efficient creative industrial city, it is more instrumental and determined by the 'spirit' of capitalism. In such a city, the creative industries, particularly advertising, marketing, media and those involved in the formulation of images, are paramount as they are the ones who seek out the new, it is they who are at the 'leading edge' of capitalistic development. Chapter 2 discussed how city branding is homogenising urban visions globally, and so within the creative industrial city urban subversions are used to promote the city to the creative class as the 'most creative' place to live, work and play. The subversive characteristics of the practices are redacted as soon as they happen, as they instantly become part of the city's functionality of branding. Moreover, the city begins to deliberately force subversion (through creating 'gaps' in the system) so it can recognise it even quicker (as it knows when and where it will materialise). The neoliberalised assemblage of the creative industrial city can enact *and redact* state provision where necessary (such as the introduction of austerity politics) in order to force urbanites into creative reappropriation to address the 'lack'. In effect, desire-production can be channelled to produce the new forms that will give the city an even greater competitive edge.

Such a system can be glimpsed today. The rise of Tactical Urbanism as an identifiable and concretely defined policy for entrepreneurial urban managers exemplifies how disparate forms of urban subversion are being channelled into creating new forms of replicable industrial creative urbanism (in the Lefebvrian sense). I have outlined the details of Tactical Urbanism elsewhere (Mould 2014b) but it is worth revisiting the reasons as to why it is inculcating a creative industrial city. Tactical Urbanism is a phrase articulated by Mike Lyndon, who is part of the Streets Plan Collaborative, an institution that seeks to 'create high-quality public spaces, believing that the key to reversing the harmful effects of suburban sprawl is to promote compact, walkable, mixed-use neighborhoods' (Streets Plan Collaborative, n.d.). Lyndon *et al.* penned an online publication called *Tactical Urbanism: Short Term Action, Long Term Change* in 2011 (with a second volume in 2012). It states:

> Improving the liveability of our towns and cities commonly starts at the street, block or building scale. While larger-scale efforts do have their place, incremental, small-scale improvements are increasingly seen as a way to stage more substantial investments ... Sometimes sanctioned, sometimes not, these actions are commonly referred to as 'guerilla urbanism', 'pop-up urbanism', 'city repair' or 'DIY urbanism'. For the moment, we like 'Tactical Urbanism'. (Lyndon *et al.* 2012: 1)

The publication offers ways of performing Tactical Urbanism and what potential limitations there can be from urban officialdom. Examples they offer include turning parking spots for automobiles into temporary parks by placing potted plants, deck chairs, tables and even laying down turf (called 'parklets'), putting chairs and tables down in plazas, guerrilla gardening, yarn-bombing and generally those activities undertaken by people who are trying to be active urban citizens rather than passive consumers of the spectacle. Tactical Urbanism is inculcating an agenda from a variety of 'everyday' subversive activities. The second volume insists that Tactical Urbanism is not new, citing *Les Bouquinistes* in Paris in the 1500s as the first informal 'pop-up' shops and therefore the world's first example of Tactical Urbanism (this is despite market stalls and mobile street vendors being common throughout antiquity). But the inclusion of 'pop-ups' as part of an urban strategy of citizenry is particular telling given the very reasoning 'pop-ups' exist in the first instance.

A 'pop-up' institution, be it a cafe, library, market stall, art exhibition or shop, has become very popular within urban developmental discourse of late (Rosenberg 2011). This is where temporary stands are erected (usually) in public spaces that have a commercial usage for a short period of time. In the wake of the recent financial crisis, many projects have been suspended due to lack of finance. This has had a profound impact upon the retail sector, with (for example) one in ten shops in the UK vacant in 2012 (BRC 2012). This dereliction in the urban landscape though is a direct consequence of state-led strategies of austerity and reduction in public spending in urban councils, which has forced more entrepreneurial strategies to be formulated (see Peck *et al.* 2013). In other words, it can be seen as the creation of a 'gap' in the system, one that the creative entrepreneurs will fill. One such initiative is to encourage small firms, individual freelancers and business entrepreneurs to set up temporarily (or to 'pop up') in vacant shops or to occupy the site that was previously to be built upon permanently by offering short-term leases. This broadly appeals as the private landowners receive an income of sorts, and the city reduces its dereliction rates, if only temporarily. And, as is putative within contemporary urban practice, this process has of course

Figure 8.1 Boxpark in Shoreditch, a pop-up mall made from shipping containers.
Source: Author photo, 7 February 2013.

been given a name, 'pop-up'. As such, the term 'pop-up' has gained significant traction in its own right in recent years, entering the business vernacular and mainstream media as a viable model for operation. Numerous 'pop-up' cafes and shops have occurred in vacant lots, as well as being associated with festivals and events. A famous example can be found in Shoreditch in London. 'Boxpark', as it is known, promotes itself as the 'world's first pop-up mall' and is created out of old shipping containers (see Figure 8.1).

Boxpark was erected on a site that was earmarked for real estate development, but the plans were halted owing to the financial crisis (Iossifova 2013). The park was given a short-term lease of four years and opened in 2011. Its success to date has already seen the owner potentially apply for a longer lease, which does, however, begin to problematise the temporary nature of its own existence and subsequent success. The economic success and global fame of Boxpark along with the explosion of the pop-up phenomena has been spectacularly rapid given their very agile nature and their alignment with a style of urban development that is focused on the Floridian notion of 'cool'. This is indicative of the economic restructuring that is inherent in a new agile and creative state of urban development, brought about through an enforced climate of austerity that has been enacted in response to the recent global economic recession. There are countless other examples of how the term 'pop-up' has been used to advertise and champion temporariness and glorify the precarious and peripatetic nature of this style of urban commercial and retail activity. It exemplifies the hegemonic co-option (through linguistic neutralisation of using a 'cool' and apolitical term like 'pop-up') of community-orientated activities that originate through the financial necessity of individuals rather than the business strategies of larger firms and consumption processes. Moreover, the broader remit of Tactical Urbanism is to 'become something larger' (Mould 2014b), to be part of the urban development narrative. It represents how capitalism has created a gap in the system (via austerity politics), waited for people to fill that gap with (albeit temporary) subversion and then instantly co-opt such activity as part of a narrative of the Creative City.

So the creative industrial city absorbs urban subversion into its reproductivity. It neutralises the subversive and antagonistic politics of desire-production before they have the oxygen to formulate, channelling it instead into the development of the neoliberal assemblage of Creative City development – another example of which we saw with Guetta's 'street art' in *Exit Through the Gift Shop* (2010) detailed in Chapter 6. Hence 'subversion' is nigh-on impossible in the creative industrial city because it does not have the time or resources to collectivise and form resistive functionalities. It becomes simply a mechanism of revealing newer ways of maintaining profit margins. The resistive qualities of subversion are neutralised and whitewashed and instead their novelty is celebrated as a 'place-making' device that enlivens sterile and/or derelict places. In the creative industrial city, *true* creativity has become completely industrialised.

The spectacularised city

Debord's (1967: thesis 6) articulation of the spectacle argues that 'it is the omnipresent celebration of a choice already made in the sphere of production, and the consummate result of that choice.' As detailed in Chapter 5, the spectacle was for Debord the way in which society was organising itself. He argued that 'commodities are now all there is to see; the world we see is the world of commodity' (ibid.: thesis 42). In other words, the social world that Debord sought to critique was one formulated upon the continual representation of commodities through signs, symbols and mediated form. In Chapter 2, the role of advertising, branding and PR in urban governments was highlighted as formulating a city that is packaged, sold and circulated globally (as was the case with Sydney's aggressive marketing campaign 'Sydnicity'). The city itself has become commodity. Additionally, if we rehearse what Baudrillard (1996) noted about the difference between the function and functionality of objects (outlined in Chapter 6), we see that without attempting to desire-produce new alternative functions of urban objects, then the commodification of those objects will continue and intensify ad infinitum. The 'system of signs' that is continually constructed, manipulated and configured around urban technologies produces a city that is forever commodified, mediated and spectacular, something which Berardi (2012) has articulated as 'semiocapitalism'. The images of the city are *re*presented in the city, and the images we have of cities in our minds become the sole experience of the city. We enter a hyperreal urban.

Las Vegas, one of the most mediated, neon-riddled and ostentatious cityscapes on the planet, has often been the subject of such accusation, with Baudrillard (1994: 91–2) offering the most (in)famous description:

> When one sees Las Vegas rise whole from the desert in the radiance of advertising at dusk, and return to the desert when dawn breaks, one sees that advertising is not what brightens or decorates the walls, it is what effaces the walls, effaces the streets, the facades, and all the architecture, effaces any support and any depth, and that it is this liquidation, this reabsorption of everything into the surface (whatever signs circulate there) that plunges us into this stupefied, hyperreal euphoria that we would not exchange for anything else, and that is the empty and inescapable form of seduction.

The city of Las Vegas is a city of signs, perpetual light, branding and advertising surfaces. The city is a machine of hyperreality, defenestrating any experiential reality instantly from its surfaces. It is a complete representation of itself – it is the Debordian spectacularised city.

Las Vegas though is the scourge of the Creative City. It is often derided in the literature as having 'low-quality growth' (Florida 2005: 24) being as that growth, while rapid, is based on service class work that, as we saw in Chapter 3, is there solely to prop up the creative class. Such classification though belies the inherent spectacularised nature of work in the hyperreal, mediated city. All surplus value that is generated through work goes into feeding the mediation of the city. Inauthentic smiles, welcome teams, dos and don'ts of customer service, feedback systems, automated sincerity, even work in the spectacularised city is commodified. Moreover, in Chapter 4, we saw how Media Cities are becoming increasingly popular as a Creative City policy vehicle. Salford, Sydney, Dubai, Seoul, Copenhagen, Abu Dhabi, Helsinki, Shanghai (and many others that I have not had the opportunity to visit yet) have all built large-scale, privately managed citadels to creative consumerism. Figures 4.6 and 4.7 exemplify just two instances of the mediated architecture that these places engender. Large video walls that can advertise and mediate whatever content is needed, spewing light, information and subjectivities onto the concrete below attacking the senses. McQuire (2008) has argued that the increase in the visualisation of the public sphere has reduced the ability to engage in dérive, to be a flâneur. He argues that 'as the rhythms of both economic exchange and social life accelerate, the prospects for *flânerie* diminish' (McQuire 2008: 133). The mediation of the urban realm (both public and private) is becoming ever more saturated. Technological innovation within large display screens means that they are more and more prominent. Subway trains, lifts, clothes shop changing rooms and even in public conveniences, media screens will be wherever eyes will linger. The mediation of the city's surface in the spectacularised city is total, and so there is no experience left to discover through flânerie. We become *badauds*, gawkers that have become lost in the crowd, camouflaged into a city whose reality is one that is completely made up of signs.

However, as this book has argued, urban subversion has the ability to 'see through' the 'system of disavowal, lack and camouflage' (Baudrillard, 1996: 69) and so can begin to disrupt the continual representation of realities, reintroducing experiences into the urban realm. Even within Las Vegas, the hyperreal city par excellence, we can see subversion. O'Brien (2007) ethnographically documents the homeless community living in the storm drains, surviving in the cracks of the city by feeding off the gargantuan amounts of waste that the exuberant city generates. By engaging in urban subversion, we can desire-produce alternative functions of the urban terrain and transcend functionalities. But in the spectacularised city, the process of urban subversion is itself mediated and represented. There is the continual proliferation of the mediated imagery of subversive activity. As Chapter 6 detailed, after such activity is marginalised from the Creative City, the process of subculturalisation subsequently entails the socialisation of people into 'communities of practice' (Wenger 1998). This is far more than simply like-minded people coming together to practise, it is the development of a 'way of doing things' and the creation of 'meaningful statements about the world, as well as the styles by which they express their forms of membership and their identities as members' (Wenger 1998: 83). Such meaning and expression is transmitted to other people wanting to practise their burgeoning subculture most readily in the contemporary city via technological means. Virtual dissemination via social media, online videos and photographs, discussion forums and interactive and crowd-sourced innovations have sped up the subculturalisation process. We have already seen how parkour has become a highly visual practice with a multitude of videos, TV shows and major films all depicting it. The increase in mediation of parkour has meant its 'meaning' can be disseminated to a far wider array of people than it could be without virtual and/or digital mediation. But the 'meaning' changes, shifts and transmogrifies into a symbolic representation of parkour, one that adds a further 'stratum of commodity' (Debord 1967: thesis 42) to the spectacularised city. Nowadays, parkour has become so highly visible that it is undertaken purely for the purposes of 'doing parkour'. Its intense mediation means it is beginning to represent only itself. As such, in the spectacularised city, the embodied, affective and conscious experiences of urban subversion are continually reduced until there is nothing left but representation.

Urban creative subcultures thrive on such mediation; they could not exist without it. Skateboarding (a snapshot of which can be seen in Figure 8.2), urban exploration, parkour, yarn-bombing, flashmobbing, buildering and so on have all gained global attention because of their imagery. In becoming identifiable, they first have to be mediated – it is a critical component of the collectivisation inherent to the subculturalisation process that I detailed in Chapter 6. In the spectacularised city such mediation is total and there is no political distance between the representation of the city and the representation of urban subversion – they both serve to replicate the city as commodity. They merge into a hyperreal city that is

Figure 8.2 The mediation of skateboarding.
Source: Author's photo, Venice Beach, Los Angeles, 14 April 2013.

nothing but pure spectacularised image. Moreover, such a hyperreal city leaches from us the ability to experience the urban topology. In this city, far from being able to inculcate situations and visceral experiences, the act of walking (in relation to the discussion of De Certeau (1984) in Chapter 5) is mapped and contextualised instantly. There is no experience that becomes 'hidden', all there is is the map and the plan. In this city, urban subversion cannot free itself from pure representation and there is nothing beyond the functionality of signs to experience. There is no desire-production because there are no other functions to realise other than those that feed the continuation of the spectacle. In this city, urban subversion feeds the spectacularised city.

The creatively activist city

The Creative City continually feeds on the appropriation of urban subversion and the co-option of creative urban subcultures; the capitalistic spirit that Boltanski and Chiapello (2005) have examined in detail accounts for this. However, a city can exist where resistance to such appropriation can thrive, where antagonistic practices are rife. Within such a city, the capitalistic mechanisms of neutralising and whitewashing subversion are challenged continually, most readily through an engagement with artistic and creative practices. Producing art therefore becomes a political act. As Hawkins (2014: 100) has eloquently articulated:

> The social problems of the city [are] the very conditions of art's existence, and, in reaching out to new audiences, artworks began to constitute new publics through which to engage and debate issues of social justice in the city. Here art comes to intervene in debates about the production of urban space, and what is more, forms a site for the production of politics, requiring that we consider it in terms of human and nonhuman relations, both antagonistic and cohesive.

So in the creatively activist city, the 'production of politics' as artistic is total, hence the issues of social justice in the city, the constitution of urban space and the iteration of the social problems of the Creative City are tackled head-on through creative artistic practices. This city is full of cultural and creative clashes, with politics and counter-politics conveyed with every act of subversion. It is difficult not to encounter such messages within the contemporary Creative City. This book has exemplified urban subversion in its most recognisable form (i.e. creative urban subcultures) but artistic urban activism is rife throughout the urban quotidian (not least with the poignant example of Soldier Matt in Chapter 6). Only the most sterile, policed, privatised, securitised and micro-managed (such as those discussed in Chapter 4) do not have some sort of artistic defacement.

But in the creatively activist city, urban subversion occurs continually against the Creative City. There is continual creativity in reaction to the mundane 'creative' practices of the Creative City. Alternative functions of the urban topology are constantly realised and desire-production perpetually sparks new forms, knowledges, subjectivities and manifestations of becoming that build up to a point of contestation and antagonism toward the prevailing hegemony. However, the activist practices and the ossified instances of them are then challenged the moment they arise. Subcultures struggle to come into being because their gendered, class and ablist normalising tendencies are contested pervasively, meaning that communities of practice are rarely able to form. So instead of subcultural creation, there is constant critique, questioning and becoming-minor. In such a city social groupings only loosely take hold, as there is always instability and insatiability is unchecked at the margins. A *potential* result could be Durkheimian anomie, as individuals conducting subversion fail to connect and socialise. Such fatalistic tendencies, however, render the activist potential of subversion mute. Hence in the creatively activist city, such collectivisation only occurs so as to form the politics of subversion, and then continues to flee once the message has been put across. It is a city of De Certeauian tactics. Creative activist practices take place momentarily, but they disband and disengage and burrow further into the strategy of the Creative City. They form a Deleuzo-Guattarian Body without Organs (BwO);

> Thus the BwO is never yours or mine. It is always *a* body. It is no more projective than it is regressive. It is an involution, but always a contemporary,

creative involution. The organs distribute themselves on the BwO, but they distribute themselves independently of the form of the organism; forms become contingent, organs are no longer anything more than intensities that are produced, flows, thresholds and gradients. (Deleuze and Guattari 1987: 164, original emphasis)

In other words, creative activist practices are *involutions*, organs that appear when necessary but then afterwards dissipate. They flow/flee continually. Moreover, a BwO is precisely that, *without organ*isation. It has no centralising force that can ossify into hegemony. Creative and artistic activist practices do not linger long enough to gain the traction that creates something to be capitalised. There is enough collective social action to create and articulate the artistic critique but that is all. They do not allow the strategy to take hold; they evade recapture.

We do not have to look far for examples of such activity. Routledge (2010) documents the activities of the Clandestine Insurgent Rebel Clown Army (CIRCA), who partake in 'ethical spectacles' which are 'participatory, open-ended, and playful urban transformations whereby the politics desired is embodied in the means of the spectacle' (Routledge 2010: 431). They are interventionist, subversive and, as the name suggests, clandestine in their operations. Famous for the intervention at the G8 summit at Gleneagles in Scotland in 2005, CIRCA have gained notoriety for their 'cultural activism' that is based on the 'playful' activities of circus clowns, which is what they are often dress up as to conduct their activist activities. Other groups such as the Yes Men also perform spectacular interventions that are political and have anti-commercialist, anti-globalisation, anti-war and anti-hegemonic politics (Mouffe 20213). Other groups include Brandalism UK, who engage in 'subvertising', which is 'throwing small visual or verbal spanners in the work of advertising' (Bearder 2012: 6). However, the very fact that they can be identified as a group renders them elusive in the creatively activist city. Subversive activity in this city never gets that far; collective action is undertaken, just enough to 'get the job done'. It is project-based, it is never sustained.

Such a city crackles with conflicting voices. Hegemony in the creatively activist city is constantly attempting to marginalise activities, but as they do not coagulate into identifiable groups (subcultures or activist collectives), these efforts are futile. Each subsequent artistic activist project is creative enough so as to elude attempts to guard against it. Each new realisation of an alternative possible function defies attempts to restrict it precisely because its manifestation cannot be known or even conceived of by the Creative City paradigm. The city is full of political slogans, artistic interventions and temporary, fleeting but concentrated and intense occurrences of counter-cultural activity. Yet, the heterotopic politics of the city are not stabilised or coexistent. Counter-counter-cultural messages are created, activist activity itself is critiqued and then critiqued further, and there is a constant conflict within and between the differing political heteroglossia. The city's layering of texts creates an undecipherable palimpsest. The multiplicity of voices and reappropriated topologies continually assert themselves, with the resultant 'noise' becoming all consuming. This is what Mouffe (2013) has theorised as an agonistic ontology. For this approach, 'public space is where conflicting points of view are confronted without any possibility of a final reconciliation' (Mouffe 2013: 92). Far from trying to reach a consensus, the creatively activist city strives for messiness and the complete deconstruction of hegemonic subjectivities. The aim of these critical subversions is to give a voice to those who are silenced by the existing power structures. However, it is important to realise, that 'critical artistic practices, in this view, do not aspire to lift a supposedly false consciousness so as to reveal the "true identity"' (Mouffe 2013: 93). There is no consensual reality to reveal, no utopian discourse that is waiting to be 'unearthed'. Political motivations cannot be coerced to a more appropriate alternative way of thinking; there is no rationalism within the creatively activist city that ascribes a true identity to be revealed. Rather, 'the social agent [is inscribed] in a set of practices that will mobilise its affects in a way that disarticulates the framework in which the dominant process of identification takes place' (ibid.). In other words, urban subversion in the creatively activist city is not about revealing a hidden utopia – it *is* the utopia (or perhaps heterotopia (Foucault 2008 [1967])). Artistic practices are solely for the purpose of questioning the systems in which it operates, and so there is a form of nihilism in the creatively activist city that is pervasive and difficult to extricate from. Urban subversion in the creatively activist city does not produce consensus, nor does it fight for a singular cause. It constantly fans the flames of agonism, creating discontent and deconstructing subjectivities wherever it is practised.

The socially creative city

It has been articulated throughout Part II of the book that subcultural activity is predicated upon socialised activity. The coming together of like-minded individuals to practise their (marginalised) activity is a key phase in how it becomes identifiable. Such collective action (including online, virtual and mediated disseminations), according to Deleuze and Guattari (1987), is how we can aid escape from the apparatus of capture. Collectivisation, strength in numbers, is the best course of action in order to resist co-option from the Creative City. To resist together is to resist better. Deleuze and Guattari also postulate that the BwO is the most effective organisation in resisting, as it is non-hierarchical, without a centre and able to 'grow' from any part. There is no rupturing the BwO, necessary 'organs' are produced whenever and wherever they are needed. However, as we saw in the creatively activist city, such organisation leads to full-on agonism and the constant deconstruction of functionalities, signs and subjectivities. But, as we saw in the spectacularised city, collectivisation of subversive activity can lead to the complete mediation and loss of experience and performativity, and all that exists is total representation.

But if collectivised creative action can eschew both these scenarios with a resource that denounces hegemony and representation in all its forms, but at the same time can also produce meaningful new subjectivities that can impact upon the city beyond putative antagonism, then a socially creative city comes into focus, one that relies upon social bonds, collective creativity and community action. Such a city is not unlike Constant's New Babylon (detailed in Chapter 5), in that it relies on subversion that is purely collective, and while not aiming to counter, resist and displace the prevailing Creative City, there is a set of 'rights' or a Lefebvrian 'contract' in place. Lefebvre (2009: 134) has argued that 'we have seen exchange and the commodity, capitalism and statism "assimilate" actions and ideas that seemed essentially subversive.' And so, because of this, Elden (2004: 231) describes how Lefebvre sees that the current 'citizenship contract' between people and the state is unbalanced and should be:

> ... broadened to include a right to information; the right to freedom of expression; the right to culture (to enjoy art and explore the world); the right to identity and equality within an understanding of difference; the right to the city; the right to public services and the right to *autogestion*.

Within the socially creative city, it is the right to the city and the right to *autogestion* which are most critical. Autogestion, Lefebvre posits, is when groups of people who are discontented and/or deprived work together to actively manage their own system of affairs. Be they workers, students, a residential community or the 99 per cent, autogestion is the process of self-management, taking control of their own existence, undertaken in response to the appropriating forces of hegemony. Indeed, he has noted that 'each time [a social] group forces itself not only to understand but to master its own conditions of existence, autogestion is occurring' (Lefebvre 2009: 135). The occurrence of autogestion is not, however, easy. Purcell (2013: 40) has argued that 'because it is by definition a process whereby people govern themselves, autogestion cannot be brought about intentionally by a vanguard. It cannot be organised by a cadre of activists, nor can it be prompted by party leadership.' Autogestion is a collective 'spirit' that matches the 'spirit' of capitalism that Boltanski and Chiapello (2005) articulate so explicitly. In the socially creative city then, autogestion occurs constantly and people are able to self-manage their own creativity and use it for their own needs and goals. They channel desire-production just so much as to create functionalities that serve the needs of the community, but beyond that, desire-production is unchecked, free to create and explore alternative functions ad infinitum.

Again, there are examples that we can draw upon. Christiania in Copenhagen is an area of the city that is legally and politically ambiguous from the viewpoint of the municipality, yet it is a thriving creative community. Started in 1971 (with very much a nod to unitary urbanism and the Situationists), it is a commune of sorts that has its own special laws distinct from the surrounding city of Copenhagen. The area is home to approximately 850 people, but none of them 'own' their houses legally. There are no police and the only laws are no running, no hard drugs, no violence and no photographs. There are no hotels, no marketing or branding initiatives and the businesses that are situated there are small-scale selling produce made, reared and sourced within the commune. Moreover, Christiania is an area rife with creative expression;

> Every small corner of public space hosts a sculpture, a painting, a wrought-iron statue or an installation; every house is partially or totally self-constructed in a curious and colourful way. (Vanolo 2013: 1791)

The community is fully integrated and in visiting the area, it is difficult not to feel a responsibility to maintain the community by adhering to their rules. But to do so is through an emancipatory affect, not from a fear of oppression from the processes of the privatisation of space. In other words, the locality encourages you to collaborate in their social community through those tacit qualities of serenity, joviality and collective endeavour. The sociality of the area therefore is highly evident not just in the physical connections of people, but also through the qualitatively pervasive and affective aura of collaborative existence. Christiania is a place of pervasive autogestion, a socially creative city in microcosm. The intense artistic creativity is not riddled with counter-hegemonic narratives but pure expressive freedom. Given that the community self-manages up to the point where needs are met, the surplus of energy, labour and value goes not on producing profit, hierarchy and inequality, but is free to desire-produce. This manifests in a *purer* form of creativity that is often surreal, but only as it is not adhering to the normative conditions of the Creative City (and the subcultures it produces). Local farming, healthcare and even informal educational institutions have been set up. It is, of course, not without its problems. There are sparks of violence, attempts at hierarchy and coercion, land disputes and disagreements (the inhabitants still have to pay fees to the original land owners, the Danish Ministry of Defence). The area is under constant threat from closure as the legal disputes with the city of Copenhagen continue. It is, after all, a city within the Creative City. Despite this, 'Christiana is a laboratory for resistance and creation and, given its desire not to be co-opted, intrinsically an exceptional, limited and ephemeral space' (Vanolo 2013: 1798).

So, as in Christiania, urban subversion in the socially creative city comes from the desire for collaborative self-management. It is subverting the idea that existence, survival and creativity is dependent upon an externalised power; it is the realisation that desire-production can be free from such forces if the social will allows it to be. By creating just enough stasis and normativity to offer what is needed, the remaining effort and resources that subversion affords creates a plethora of creative expressions throughout the city. But in the socially creative city, the threat of co-option is always there. Like Christiania's constant fight for its own survival, or the campaign to save the undercroft from the Festival Wing development (detailed in Chapter 7), the Creative City is always looking to reappropriate desire-production for the establishment of its own status quo. This is why it is important in the socially creative city that such instances of complete autogestion are encouraged not just from within, but externally as well.

In the socially creative city, not only are pockets of autogestion visible, but they are encouraged to continue in the face of uncertainty.

Purcell (2013) eloquently guards against a city in which hegemonic forces subdue the masses into passivity. Discussing Calvino's *Invisible Cities*, he argues that we need to follow his example:

> Seek and learn to recognise those things, both in society and in yourself, that in the midst of inferno are not inferno, and nurture those things, give them space and help them flourish on their own terms. (Purcell 2013: 151)

Therefore, in a city of inferno (capitalistic appropriation), there is a need to recognise the not-inferno and help them grow how they want to. A socially creative city therefore encourages its pockets of not-inferno to grow, to self-manage and to create new forms that do not feed the Creative City. But to engage in such activity requires another resource, namely an *ethics*. In the socially creative city, subversion is not only collective and self-managing, it is ethical. Not ethical in a way that is currently understood within capitalist discourse (ethical consumption or the fight for human rights for example), but based on an ethics that Badiou details – an ethics that forces us to remain 'truthful' to an event.

Badiou (2001, 2005) speaks of an 'event' that ruptures the state of the situation or the status quo. Such events are for Badiou extremely rare because they completely change everything about the state of that prevailing situation: the way it works, operates, functions, looks and feels. The event is so new, so novel, that it exposes the *void* in that situation. Up until the event, the state of any situation contains unspoken or unarticulated truths that are hidden and unknown. What an event does is rupture the state of the situation so much that all these hidden 'truths' become evident all at once (although some are immediately closed down again and remain confined to the virtual). For Badiou, the event therefore is the ultimate act of creativity. Moreover, events can only exist in four realms – art, science, politics and love. In each Badiou gives examples of an event. For art there is Hayden creating what could be described as classical music, in science there is Galileo's creation of a new physics, in politics there is the French Revolution, and in love there is the encounter between Héloïse and Abelard. In each case, the event rewrites the rules of the entire situation in totality. There exists no singularity that is not affected by the event. Badiou goes on to suggest, however, that truth is not somehow 'revealed' by the event, but truth is continually processed by fidelity *to* the event, which is why it is often referred to as a *truth-event*. Indeed he argues a truth

> ... is thus an *immanent break*. 'Immanent' because a truth proceeds *in* the situation, and nowhere else – there is no heaven of truths. 'Break' because what enables the truth-process – the event – meant nothing according to the prevailing language and established knowledge of the situation. (Badiou 2001: 42–3, original emphasis)

By being faithful to that event in the face of 'evil' (which Badiou cites as a trium-virate of simulacra and terror, betrayal and the unnameable), that is how 'truth' is made manifest, it is how we become human subjects. So, the initial rupture of an event is all encompassing and completely changes the 'state of play' in the situation. Moreover, being faithful to the practices of that event (through practice, ontology and belief) is how truth is made, not revealed. It is also important to note that Badiou talks of truth as being different from knowledge which changes, can be manipulated and transformed to effect discord to the fidelity. Knowledge is opinion, it is not truth. It is important to human communication, but it obscures truth and constructs evil and therefore is to be avoided. Being *ethical* therefore, for Badiou, is being faithful to the truth-event. It is hence reflexive and time-place dependent. There is no universal ethics to be adhered to, no universal global version of 'rights' that should be defended violently if needed. Such a version of ethics, Badiou argues masks neoliberal and/or imperialistic hegemonies. Rather, ethics should be a suite of resources that are at hand to people who need help to remain faithful to a truth-event, to overcome the evils that he outlines. Of particular interest to the socially creative city is an ethics that encourages persistence in the face of exhaustion or disillusionment:

> A crisis of fidelity is always what puts to the test, following the collapse of an image, the sole maxim of consistency (and thus ethics): 'Keep going!' Keep going when you have lost the thread, when you no longer feel 'caught up' in the process, when the event itself has become obscure, when its name is lost or when it seems that it may have named a mistake, if not simulacrum. (Badiou 2001: 78–9)

For Badiou then, ethics is a way of remaining truthful in the face of adversity (something which I would argue can be ascribed to the LLSB campaign outlined in Chapter 7). In the socially creative city then, all inhabitants need to be ethical to the truth event (which for the socially creative city would have looked very similar to what occurred in Paris in May 1968, providing as it did a new vocabulary of urban subversion, as detailed in Chapter 5). By being ethical, inhabitants are encouraging the not-inferno to flourish, not only in the face of co-option and appropriation, but also of personal evils such as self-doubt, exhaustion and disenfranchisement. In the socially creative city, we are all instigators of subversion as we are all encouraging the creation of a new type of creative expression free from co-option by the Creative City. The socially creative city is the closest we can come to *the* creative city.

Summary

The four scenarios given above clearly offer a dystopian/heterotopian/utopian vision of what urban characteristics come into focus with differing intensities of urban subversion. These are clearly not meant as predictions or manifestos of urban futures, but as theoretical signposts of what kind of activities and results emerge when thinking about urban subversion and the realisation of a creative city. The contemporary Creative City contains elements of all four of these (and many more). They ebb and flow temporally and spatially between these ideas. Walking the short distance from the undercroft skate spot, up the metal stairs to the outside foyer of the Queen Elizabeth Hall will take you from a socially creative city to a spectacularised city. Standing in Leake Street over the course of a few years, you would have seen it change from a creatively activist city to a creative industrialised city. Walking along the streets of Dubai Media City, you experience nothing but the creative industrialised city, yet embarking upon dérive through the streets of Seoul or Shanghai and you encounter all four of these cities all at once.

The contemporary Creative City, as this book has shown, is full of injustices, inequalities and gentrification, yet it is also full of potentialities that are there to be explored by engaging in urban subversion in all its forms. Being creative in the Creative City therefore means constantly searching for these potentialities, seeking out new functions and shunning any attempts at co-option, appropriation, representation, contextualisation, capture and identification. To create is to constantly 'veer away from representation' (Balanuye 2010: 161), for you cannot capture something that is constantly becoming-minor. Being forever critical and engaging in critique is the act of creation for, as Deleuze (1994: 176) tells us, 'the conditions of a true critique and a true creation are the same'.

To counter the deleteriousness of the Creative City requires everyone to adopt such a critique. Not a criticality that simply feeds back into the status quo by offering alternatives that already exist, but a critique that forces the creation of the new. This requires everyone, it requires collective creativity and is not limited to the purview of subcultures. Realising a creative city should encourage interactivity, the breaking down of socialised barriers and their restrictive hegemonic forces. This book has focused on cities of the Global North, as that is where my research took me. London, New York, Dubai, Sydney, Shanghai, Seoul, Copenhagen and the rest of the cities that I explored for this book – they all have different creativities to explore, yet further exploration of the cities of Latin America, Sub-Saharan Africa and South East Asia will offer even further difference, further functions to be explored, further subjectivities, cultures, hegemonies to be subverted. This book is not an attempt to eulogise the subversive tactics of the Global North. Indeed, there are many high-profile examples of urban subversion in the Global South that have been heralded as progressive manifestations of community and collective subversion, autogestion and ethics. The 'La Torre de David' in Caracas stands out in this regard (McGuirk 2014), but there are many, many others. There tends to be a Westernised epistemology to creative urban subcultures, as they are mobilised from the Global North to other parts of the world. For example, parkour is practised readily in Gaza, but those who partake are often threatened by the intense militarised violence that characterises the volatility of the region. The young traceurs from Gaza, however, through their agency in subverting the prevailing mediated narratives that would see them labelled as 'victims, ideologues or fundamentalists' (Thorpe and Ahmad 2013: 21), are creating new emancipatory subjectivities that are effecting real change in their city.

Moreover, this book has talked of only really a handful of urban creative subcultures and their subversive qualities. Skateboarding, parkour and urban exploration have been the main targets given that these are the activities that I most readily engaged with. Their normalising biases have been well documented and challenged, yet further research could be conducted on other such activities that go about tackling these biases to greater degrees. Yarn-bombing, for example, has its roots in the feminist activism of the 1970s and 1980s (Wallace 2013). Adams and Hardman (2013) also argue that the micro-politics of guerrilla gardening is not really a celebration of subversion and transgression, but a socially responsive activity that includes people and places that are perhaps marginal to mainstream visions of urban subversive activity (the elderly and younger children for example). Urban subversion undertaken by those whose bodies do not conform to the hegemonic city is also increasing: 'Adaptive Movement' is a parkour class started in Houston for those with physical impairments while Skateistan in Cambodia encourages skateboarding for people who are normally confined to a wheelchair. But more than simply retrofitting existing creative subcultures to cater for people who would not ordinarily be able to partake in urban subversion, becoming-minor allows for those labelled by society as disabled to reconfigure the city for their own needs and desires. The now famous example of Samuel Nobile de Oliveira, a disabled man who took it upon himself to build a ramp to gain access to Juína's municipal health building stands out in this regard. His desire-production embarrassed the city into acting, thereby creating that little bit extra social justice in the city. Disability art is often reactive and critiquing the ablist tendencies of the city and society more generally, as is the case with the Deaf artist Christine Sun Kim who is 'unlearning sound etiquette' and creating new sounds and new subjectivities in relation to those sounds through the subversion of the 'traditional' rules of sound etiquette and how hearing people react to sound. Through realising new functions of aural specificities, she is becoming-minor and exploring new subjectivities that emanate from the disruption of traditional 'rules' of sound and reaction to it. The soundscapes of the Creative City are critical to how we are channelled into consuming it. By subverting common sound patterns, Kim is 'critically destabilis[ing] the assumption that to be hearing is a natural part of life for everybody' (Harold 2013: 860), and so is creating new ways of experiencing the everyday urbanity beyond the (majority) hearing world.

The Creative City paradigm is dead but still dominant. We have overseen its destruction in Part I, but there is a need to continue to resist its dominance via continual urban subversion, as Part II has articulated. We need to disrupt this process of dominance, and rescue the idea of 'being creative' from the neoliberalised, hegemonic dogmatic term that it has become. Therefore this book has hopefully shown that to be creative – *truly* creative – is to realise new functions of the city, to become-minor, to be tactical, to subvert heteronormative, class and ablist biases wherever they occur, to constantly flee from the apparatuses of capture – in sum, to engage in urban subversion. But this is a collective endeavour; we cannot do this as a set of individuals. It requires a collaborative effort that releases the socialised agency of all people to effect change in the city. We need to construct an urban fabric of creative sociality. This requires supporting each other when struggles and disenfranchisement abound. It requires us *as a totality* to shun passive consumption of the Creative City. Together, being active urban citizens *is* to be truly creative.

Epilogue
Climbing Wooly

In the Spring of 2013, I came into contact with an urban explorer in New York for an interview. A few hours later, I was with a group of people who were also novice urban explorers, climbing the stairs inside the Woolworth Building (known as 'Wooly') in Manhattan, one of the city's most iconic buildings and the world's first ever skyscraper, less than two blocks and half a mile away from where Phillipe Petit performed his tightrope walk between the Twin Towers 39 years previously. The explorer did not come with us but had given us explicit instructions on how to obtain access to the building's pinnacle. So, dressed like we were supposed to be there (in a suit and smart trousers), we sweet-talked our way past security on the ground floor (giving an excuse tried and tested by the exploration community which works every time) and took an elevator to the 22nd floor. Trying to look nonchalant and inconspicuous, I felt anything but, my eyes flitting between the strangers I had just met that I was exploring with and the other strangers in the lift who presumably had no idea (or care) about what we were doing.

We exited the lift into a corridor that looked like any other office building, with closed wooden panelled doors concealing the din of affective and financial labour. We continued to walk down the corridor with a faux-bravado that would be needed if anyone was to spot us, but with an anxiety that the door the seasoned explorer had told us to find may be locked, covered over or moved (as the remaining floors of the building above were a construction site). We turned a corner, and found the door, just as the instructions had indicated. Opening the door, we came across a metal stairwell that seemed to border the lift shaft. Craning my head up, I saw 30 floors of red metal stairs, zigzagging up into a vertical vanishing point. We puffed out our cheeks and started to climb. Wheezing and panting we continued to climb these stairs, and every so often through the walls we spotted the stagnant construction sites: wires hanging from ceilings, instructions painted on walls in fluorescent yellow paint, exposed steel girders, piles of plasterboard, bricks and wood rubble and other general construction detritus that looked like it had remained untouched for some time. (The typical view is pictured in Figure E.1.) Apparently the top 30 floors or so of the Woolworth Building had been earmarked for conversion into luxury

Figure E.1 The halted construction site of the upper floors of the Woolworth Building.

Source: Author's photo, 20 March 2013.

condominiums before the recession but work had completely halted due to lack of funds.

The final few floors were more open and we needed to find an old spiral stair-case that lead to the very pinnacle of the building. We found the staircase, barely one person's width. We navigated the stairs and came across the door to the outside. It was stiff and took brute force to open against the howling winds outside. But once the door was opened, we were rewarded with one of most recognised, most pictured, most mediated and most romanticised-about city-scapes in the world (Figure E.2) We wedged the door open to assure our even-tual descent and began to take as many photos and films as our memory cards would allow.

By operating at the boundaries of the city's institutions, we had glimpsed a De Certeauian view of Manhattan from a vantage point that few people would ever get to see, particularly since the destruction of the World Trade Center's observation decks on 11 September 2001. The view itself alone was worth the relative precari-ous and risky exploration (although no laws were broken, only open doors walked through), but the experience of the city in those brief moments was something unobtainable by 'official' means. We had created new functions, subjectivities,

Figure E.2 The New York City skyline from atop the Woolworth Building.

Source: Author's photo, 20 March 2013.

experiences and social connections that would not have been achievable via more 'official' means. A few moments of enforced silence and scampering to hide upon hearing footsteps on the metallic stairwell on the way down were the only obstacles to an otherwise exhilarating experience of urban exploration. Operating informally, but safely and socially responsibly, that experience of New York was one that would not be matched by other 'official' observation decks (proved by the rather insipid trip to the 'Top of the Rock' that evening). Such activity can never be formally recognised or legislated for (nor should it be), but is characteristic of what the essence of searching for a creative city should be.

Bibliography

Aaker, D. (2012) *Building Strong Brands*. London: Simon & Schuster.

Adams, D. and Hardman, M. (2013) 'Observing guerrillas in the wild: reinterpreting practices of urban guerrilla gardening', *Urban Studies*, 51 (6): 1103–19.

Ameel, L. and Tani, S. (2012) 'Parkour: creating loose spaces?', *Geografiska Annaler: Series B, Human Geography*, 94 (1): 17–30.

Amin, A. and Thrift, N. (2002) *Cities: Reimagining the Urban*. London: Polity Press.

Amin, A. and Thrift, N. (2007) 'Cultural-economy and cities', *Progress in Human Geography*, 31 (2): 143–61.

Anderson, C. (1974) *The Political Economy of Social Class*. New York: Prentice Hall.

Angel, J. (2011) *Ciné Parkour*. London: Julie Angel.

Atencio, M., Beal, B. and Wilson, C. (2009) 'The distinction of risk: urban skateboarding, street habitus and the construction of hierarchical gender relations', *Qualitative Research in Sport and Exercise*, 1 (1): 3–20.

Atkinson, M. (2009) 'Parkour, anarcho-environmentalism, and poiesis', *Journal of Sport and Social Issues*, 33 (2): 169–94.

Badiou, A. (2001) *Ethics: An Essay on the Understanding of Evil*. London: Verso.

Badiou, A. (2005) *Being and Event*. London: Continuum.

Bagaeen, S. (2007) 'Brand Dubai: the instant city; or the instantly recognizable city', *International Planning Studies*, 12 (2): 173–97.

Bain, A. and McLean, H. (2013) 'The artistic precariat', *Cambridge Journal of Regions, Economy and Society*, 6 (1): 93–111.

Bakhtin, M. (1981 [1935]) *Discourse in the Novel. The Dialogic Imagination: Four Essays*, trans. Michael Holquist and Caryl Emerson. Austin, TX: University of Texas Press.

Balanuye, C. (2010) 'Becoming an expression in Deleuze: two cases from Turkey (Fazil Say and Misirli Ahmet)', *Kritike*, 4 (2): 155–65.

Banks, M. and Hesmondhalgh, D. (2009) 'Looking for work in creative industries policy', *International Journal of Cultural Policy*, 15 (4): 415–30.

Baudrillard, J. (1994) *Simulacra and Simulation*. Chicago: University of Michigan Press.

Baudrillard, J. (1996) *The System of Objects*. London: Verso.

BBC (2009) 'Curve Chief Executive Steps Down'. http://news.bbc.co.uk/1/hi/england/leicestershire/8270989.stm (accessed 21 January 2013).

BBC (2011) 'About the Move'. http://www.bbc.co.uk/aboutthebbc/bbcnorth/about.shtml (accessed 17 February 2012).

Bearder, P. (2012) 'Word of the street: subvertising and rewriting the urban visual landscape with street art', in L. Bell and G. Goodwin (eds), *Writing Urban Space*. Alresford: Zero Books, pp. 6–14.

Benjamin, W. (1997) *Charles Baudelaire. A Lyric Poet in the Era of High Capitalism*. London: Verso.

Benjamin, W. (1999) *The Arcades Project*. Boston: Harvard University Press.

Bennett, A. (1999) 'Subcultures or neo-tribes? Rethinking the relationship between youth, style and musical taste', *Sociology*, 33 (3): 599–617.

Bennett, L. (2011) 'Bunkerology: case study in the theory and practice of urban exploration', *Environment and Planning D: Society and Space*, 29: 421–34.

Berardi, F. (2012) *The Uprising: On Poetry and Finance*. Los Angeles: Semiotext(e).

Bernstein, M. (2008) *All the King's Horses*. Los Angeles: Semiotext (e).

Boltanski, L. and Chiapello, E. (2005) *The New Spirit of Capitalism*. London: Verso.

Borden, I. (2001) *Skateboarding, Space and the City: Architecture and the Body*. Oxford: Berg.

Borén, T. and Young, C. (2013) 'Getting creative with the "creative city"? Towards new perspectives on creativity in urban policy', *International Journal of Urban and Regional Research*, 37 (5): 1799–815.

Brenner, N. and Theodore, N. (2002) 'Cities and the geographies of "actually existing neoliberalism"'', *Antipode*, 34 (3): 349–79.

British Retail Consortium (BRC) (2012) *21st Century High Streets: What Next for Britain's Town Centres?* London: BRC.

Brown, A., O'Connor, J. and Cohen, S. (2000) 'Local music policies within a global music industry: cultural quarters in Manchester and Sheffield', *Geoforum*, 31 (4): 437–51.

Brown, R. (2012) 'Deep Waters: UNESCO, World Heritage and Liverpool Waters' SevenStreets'. http://www.sevenstreets.com/liverpool-waters-peel-unesco-whs/ (accessed 4 September 2014).

Castells, M. and Mollenkopf, J. (1991) 'Introduction', in M. Castells and J. Mollenkopf (eds), *Dual City: Restructuring New York*. New York: Russell Sage Foundation, pp. 3–22.

Catsoulis, J. (2010) 'On the Street, at the Corner of Art and Trash', *New York Times Online*, 15 April. http://www.nytimes.com/2010/04/16/movies/16exit.html?ref=moviesand_r=0 (accessed 5 September 2014).

Chow, B. (2010) 'Parkour and the critique of ideology: turn-vaulting the fortresses of the city', *Journal of Dance and Somatic Practices*, 2 (2): 143–54.

Christophers, B. (2008) 'The BBC, the creative class, and neoliberal urbanism in the north of England', *Environment and Planning A*, 40 (10): 2313–29.

City of Sydney (2013a) 'Sydney2030/Green/Global/Connected'. http://www.sydney2030.com.au/learn-in-2030/art-and-culture/creative-city (accessed 3 September 2014).

City of Sydney (2013b) 'Our Global City'. http://www.cityofsydney.nsw.gov.au/learn/research-and-statistics/the-city-at-a-glance/global-sydney/our-global-city (accessed 4 September 2014).

City of Sydney (2014) 'Creative City: Draft Cultural Policy and Action Plan 2014–2024'. http://www.cityofsydney.nsw.gov.au/__data/assets/pdf_file/0011/213986/Cultural-Policy-and-Action-Plan-2014-2024.pdf (accessed 19 November 2014).

City of Sydney (n.d.) 'Cultural Ribbon'. http://www.sydney2030.com.au/learn-in-2030/art-and-culture/cultural-ribbon (accessed 3 September 2014).

Clement, M. (2012) 'Rage against the market: Bristol's Tesco riot', *Race and Class*, 53 (3): 81–90.

Comunian, R. and Mould, O. (2014) 'The weakest link: creative industries, flagship cultural projects and regeneration', *City, Culture and Society*, 5 (2): 65–74.

Coverley, M. (2012) *Psychogeography*. London: Oldcastle Books.

Cox, T. and O'Brien, D. (2012) 'The "scouse wedding" and other myths: reflections on the evolution of a "Liverpool model" for culture-led urban regeneration', *Cultural Trends*, 21 (2): 93–101.

Cresswell, T. (1996) In Place/Out of Place: Geography, Ideology, and Transgression. Minneapolis, MN: University of Minnesota Press.

Cresswell, T. (2013) *Place: A Short Introduction*. London: Wiley-Blackwell.

Daskalaki, M. and Mould, O. (2013) 'Beyond urban subcultures: urban subversions as rhizomatic social formations', *International Journal of Urban and Regional Research*, 37 (1): 1–18.

Davies, W. (2014) *The Limits of Neoliberalism: Authority, Sovereignty and the Logic of Competition*. Sage, London.

Davis, M. (1990) *City of Quartz: Excavating the Future in Los Angeles*. London: Verso.

Davis, M. (2006) 'Fear and money in Dubai', *New Left Review*, 41: 47.

DCMS (1998) *Creative Industries Mapping Document*. London: Department of Culture, Media and Sport.

DCMS (2001) *Creative Industries Mapping Document*. London: Department of Culture, Media and Sport.

De Certeau, M. (1984) *The Practice of Everyday Life*. London: University of California Press.

Debord, G. (1983 [1967]) *The Society of the Spectacle*. London: Rebel Press.

Deleuze, G. (1994) *Difference and Repetition*. New York: Columbia University Press.

Deleuze, G. and Guattari, F. (1987) *A Thousand Plateaus: Capitalism and Schizophrenia*. London: Continuum.

Dickens, L. (2008) 'Finders keepers': performing the street, the gallery and the spaces in-between', *Liminalities: A Journal of Performance Studies*, 4 (1): 1–30.

Dillon, S. (2007) *The Pamlipsest: Literature, Criticism, Theory*. London: Continuum.

Doel, M. (2001) '1a. Qualified quantitative geography', *Environment and Planning D: Society and Space*, 19: 555–72.

Edensor, T., Leslie, D., Millington, S. and Rantisi, N. (2010) 'Introduction: rethinking creativity: critiquing the creative class thesis', in T. Edensor, D. Leslie, S. Millington and N. Rantisi (eds), *Spaces of Vernacular Creativity Rethinking the Cultural Economy*. London: Routledge, pp. 1–16.

Edwardes, D. (2009) *The Parkour and Freerunning Handbook*. London: HarperCollins.

Elden, S. (2004) *Understanding Henri Lefebvre*. London: Bloomsbury.

Engels, F. (1987 [1844]) *The Condition of the Working Class in England*. London: Penguin.

Evans, G. (2005) 'Measure for measure: evaluating the evidence of culture's contribution to regeneration', *Urban Studies*, 42 (5/6): 959–84.

Evans, G. (2009) 'Creative cities, creative spaces and urban policy', *Urban Studies*, 46 (5–6): 1003–40.

Featherstone, M. (1998) 'The flâneur, the city and virtual public life', *Urban Studies*, 35 (5–6): 909–25.

Fischer, C. (1975) 'Towards a subcultural theory of urbanism', *American Journal of Sociology*, 80 (6): 1319–41.

Flew, T. (2012) *The Creative Industries: Culture and Policy*. London: Sage.

Florida, R. (2002) *The Rise of the Creative Class: and How It's Transforming Work, Leisure, Community and Everyday Life*. New York: Basic Books.

Florida, R. (2004) 'Revenge of the Squelchers: The Great Creative Class Debate'. CreativeClass.org.

Florida, R. (2005) *Cities and the Creative Class*. New York: Basic Books.

Florida, R. (2012a) *The Rise of the Creative Class: Revisited*. New York: Basic Books.

Florida, R. (2012b) 'What critics get wrong about creative cities', *CityLab*, 30 May. http://www.citylab.com/work/2012/05/what-critics-get-wrong-about-creative-cities/2119/ (accessed 4 September 2014).

Florida, R. (2013a) 'More losers than winners in America's new economic geography', *City Lab*, 30 January. http://www.citylab.com/work/2013/01/more-losers-winners-americas-new-economic-geography/4465/ (accessed 4 September 2014).

Florida, R. (2013b) 'Did I abandon my creative class theory? Not so fast, Joel Kotkin', *Daily Beast*, 21 March. http://www.thedailybeast.com/articles/2013/03/21/did-i-abandon-my-creative-class-theory-not-so-fast-joel-kotkin.html (accessed 4 September 2014).

Foucault, M. (2008 [1967]) 'Of other spaces', trans. L. De Cauter and M. Dehaene, in M. Dehaene and L. De Cauter (eds), *Heterotopia and the City: Public Space in a Postcivil Society*. London: Routledge, pp. 13–29.

Foucault, M. (2008 [1975]) *Discipline and Punish: The Birth of the Prison*. New York: Random House.

Frank, G. (1982) 'After Reaganomics and Thatcherism, what? From Keynesian demand management via supply-side economics to corporate state planning and 1984', *Contemporary Marxism*, 4: 18–28.

Friedmann, J. (1986) 'The world city hypothesis', *Development and Change*, 17: 69–83.

Galloway, S. and Dunlop, S. (2007) 'A critique of definitions of the cultural and creative industries in public policy', *International Journal of Cultural Policy*, 13 (1): 17–31.

Gans, H. (1999) *Popular Culture and High Culture: An Analysis and Evaluation of Taste*. New York: Basic Books.

Garrett, B. (2013) *Explore Everything: Place-hacking the City*. London: Verso.

Garrett, B. (2014) 'Undertaking recreational trespass: urban exploration and infiltration', *Transactions of the Institute of British Geographers*, 39 (1): 1–13.

Garrett, B. and Hawkins, H. (2013) 'And now for something completely different … thinking through explorer subject-bodies: a response to Mott and Roberts', *Antipode Online*. http://radicalantipode.files.wordpress.com/2013/11/garrett-and-hawkins-response.pdf (accessed 5 September 2014).

Gates, M. (2013) *Hidden Cities: Travels to the Secret Corners of the World's Great Metropolises; A Memoir of Urban Exploration*. New York: Penguin.

Geddes, P. (1915) *Cities in Evolution*. London: Williams & Norgate.

Gibson, C. and Klocker, N. (2004) 'Academic publishing as "creative" industry, and recent discourses of "creative economies": some critical reflections', *Area*, 36 (4): 423–34.

Gibson, C. and Kong, L. (2005) 'Cultural economy: a critical review', *Progress in Human Geography*, 29 (5): 541–61.

Gibson-Graham, J. K. (2006) *A Postcapitalist Politics*. Minneapolis, MN: University of Minnesota Press.

Giffinger, R., Haindlmaier, G. and Kramar, H. (2010) 'The role of rankings in growing city competition', *Urban Research and Practice*, 3 (3): 299–312.

Gilchrist, P. and Wheaton, B. (2011) 'Lifestyle sport, public policy and youth engagement: examining the emergence of parkour', *International Journal of Sport Policy and Politics*, 3 (1): 109–31.

Gill, R. and Pratt, A. (2008) 'In the social factory? Immaterial labour, precariousness, and cultural work', *Theory, Culture, and Society*, 25: 1–30.

Glaeser, E. (2004) *Review of Richard Florida's The Rise of the Creative Class*. Boston: Harvard University.

Godfrey, B. J. and Zhou, Y. (1999) 'Ranking world cities: multinational corporations and the global urban hierarchy', *Urban Geography*, 20: 268–81.

Goldsmith, B. and O'Regan, T. (2003) *Cinema Cities, Media Cities: The Contemporary International Studio Complex*. Sydney: Australian Film Commission.

Goodchild, P. (1996) *Deleuze and Guattari: An Introduction to the Politics of Desire*. London: Sage.

Graham, S. (2009) 'Cities as battlespace: the new military urbanism', *City*, 13 (4): 383–402.

Gramsci, A. (1995) *Further Selections from the Prison Notebooks*. Minneapolis, MN University of Minnesota Press.

Gratz, R. (2010) *The Battle for Gotham: New York in the Shadow of Robert Moses and Jane Jacobs*. New York: Nation Books.

Greenberg, M. (2008) *Branding New York: How a City in Crisis Was Sold to the World*. New York: Routledge.

Grodach, C. (2010) 'Beyond Bilbao: rethinking flagship cultural development and planning in three California cities', *Journal of Planning Education and Research*, 29 (3): 353–66.

Grodach, C. (2012) 'Before and after the creative city: the politics of urban cultural policy in Austin, Texas', *Journal of Urban Affairs*, 34 (1): 81–97.

Hall, P. (1966) *The World Cities*. London: Weidenfeld & Nicolson.

Harold, G. (2013) 'Reconsidering sound and the city: asserting the right to the Deaf-friendly city', *Environment and Planning D: Society and Space*, 31 (5): 846–62.

Harrison, J. (2014) 'Rethinking city-regionalism as the production of new non-state spatial strategies: the case of Peel Holdings Atlantic Gateway Strategy', *Urban Studies*, 51 (11): 2315–35.

Harvey, D. (1989) 'From managerialism to entrepreneurialism: the transformation in urban governance in late capitalism', *Geografiska Annaler. Series B, Human Geography*, 71 (1): 3–17.

Harvey, D. (2003) 'The right to the city', *International Journal of Urban and Regional Research*, 27 (4): 939–94.

Harvey, D. (2007) 'Neoliberalism as creative destruction', *Annals of the American Academy of Political and Social Science*, 610 (1): 21–44.

Harvey, D. (2009) 'Is this really the end of neoliberalism?', *Counter Punch*, 13–15 March. http://www.counterpunch.org/2009/03/13/is-this-really-the-end-of-neoliberalism/ (accessed 3 September 2014).

Hawkins, H. and Garrett, B. (2014) *For Creative Geographies*. London: Routledge.

Haywood, P. and McArdle, K. (2012) *The Secret Gardens Festival of Mass Narrative* [Show/Exhibition].

Hebdige, D. (1979) *Subculture: The Meaning of Style*. London: Methuen.

Hesmondhalgh, D. (2005) 'Media and cultural policy as public policy', *International Journal of Cultural Policy*, 11 (1): 95–109.

Hollingshead, I. (2012) 'Media City: can the BBC save Salford?', *Telegraph*. http://www.telegraph.co.uk/culture/tvandradio/bbc/9031837/Media-City-Can-the-BBC-save-Salford.html (accessed 17 February 2012).

Howells, O. (2005) 'The "creative class" and the gentrifying city: skateboarding in Philadelphia's Love Park', *Journal of Architectural Education*, 59 (2): 32–42.

Hoyman, M. and Faricy, C. (2009) 'It takes a village: a test of the creative class, social capital, and human capital theories', *Urban Affairs Review*, 44 (3): 311–33.

Hracs, B. (2009) 'Beyond Bohemia: geographies of everyday creativity for musicians in Toronto', in T. Edensor, D. Leslie, S. Millington and N. Rantisi (eds), *Spaces of Vernacular Creativity: Rethinking the Cultural Economy*. London: Routledge, pp. 75–88.

Hubbard, P. (2004) 'Revenge and injustice in the neoliberal city: uncovering masculinist agendas', *Antipode*, 36 (4): 665–86.

Hymer, S. (1972) 'The multinational corporation and the law of uneven development', in J. Bhagwati (ed.), *Economics and World Order*. London: Macmillan, pp. 113–40.

Imrie, R. (2001) 'Barriered and bounded places and the spatialities of disability', *Urban Studies*, 38 (2): 231–7.

Iossifova, D. (2013) 'Scarcity, Creativity and the Creative Industries', SCIBE Working Paper. http://www.scibe.eu/wp-content/uploads/2013/02/17-DI-scarcity-creativity-and-the-creative-industries.pdf (accessed 10 September 2013).

Iveson, K. (2012) 'Branded cities: outdoor advertising, urban governance, and the outdoor media landscape', *Antipode*, 44 (1): 151–74.

Jacobs, J. (1970) *The Economy of Cities*. London: Vintage Books.

Jensen, P. (2007) *Space for the Digital Age: Defining, Designing and Evaluating a New World Class Media Centre*. Copenhagen: Danmarks Tekniske Universitet.

Jessop, B. (1997) 'Capitalism and its future: remarks on regulation, government and governance', *Review of International Political Economy*, 4 (3): 561–81.

Kahn, M. (1995) 'A revealed preference approach to ranking city quality of life', *Journal of Urban Economics*, 38: 221–35.

Kanna, A. (2011) *Dubai, the City as Corporation*. Minneapolis, MN: University of Minnesota Press.

Karsten, L. and Pel, E. (2000) 'Skateboarders exploring urban public space: ollies, obstacles and conflicts', *Journal of Housing and the Built Environment*, 15 (4): 327–40.

Kidder, J. L. (2013) 'Parkour, masculinity, and the city', *Sociology of Sport Journal*, 30 (1): 1–23.

Kiddey, R. and Schofield, J. (2011) 'Embrace the margins: adventures in archaeology and homelessness', *Public Archaeology*, 10 (1): 4–22.

Klein, N. (2008) *The Shock Doctrine*. London: Penguin Books.

Knorr Cetina, K. (2001) 'Objectual practice', in T. Schatzki, K. Knorr Cetina and E. Savigny (eds), *The Practice Turn in Contemporary Theory*. London: Routledge, pp. 184–97.

Knox, P. (2012) 'Starchitects, starchitecture and the symbolic capital of world cities', in B. Derudder, M. Hoyler, F. Taylor and F. Witlox, *International Handbook of Globalization and World Cities*. Cheltenham: Edward Elgar, pp. 275–83.

Kong, L. (2007) 'Cultural icons and urban development in Asia: economic imperative, national identity, and global city status', *Political Geography*, 26 (4): 383–404.

Kornberger, M. (2012) 'Governing the city from planning to urban strategy', *Theory, Culture and Society*, 29 (2): 84–106.

Kotkin, J. (2013) 'Richard Florida Concedes the Limits of the Creative Class', *Daily Beast*, 20 March. http://www.thedailybeast.com/articles/2013/03/20/richard-florida-concedes-the-limits-of-the-creative-class.html (accessed 4 September 2014).

Kramer, K. and Short, J. (2011) 'Flânerie and the globalizing city', *City*, 15 (3–4): 323–42.

Kramer, R. (2010) 'Painting with permission: legal graffiti in New York City', *Ethnography*, 11 (2): 235–53.

Krätke, S. (2010) '"Creative Cities" and the rise of the dealer class: a critique of Richard Florida's approach to urban theory', *International Journal of Urban and Regional Research*, 34 (4): 835–53.

Lambeth Council (2014) *Lambeth Planning Applications Committee*, Case Number 14/01261/FUL. http://moderngov.lambeth.gov.uk/documents/s65514/04_Arches%20Waterloo%20Road.pdf (accessed 5 September 2014).

Landry, C. (2000) *The Creative City: A Toolkit for Urban Innovators*. London: Earthscan.

Landry, C. (2006) *The Art of City-making*. London: Earthscan.

Landry, C. (2011) 'Beyond the Creative City'. Talk given at PICNIC Festival, September. http://vimeo.com/30733079 (accessed 3 September 2014).

Landry, C. and Bianchini, F. (1995) *The Creative City*. London: Demos.

Langford, B. (2006) 'Seeing only corpses: vision and/of urban disaster in apocalyptic cinema', in C. Linder (eds), *Urban Spaces and Cityscapes: Perspectives from Modern and Contemporary Culture*. London: Routledge, pp. 38–48.

Lees, L. (2003a) 'Super-gentrification: the case of Brooklyn Heights, New York City', *Urban Studies*, 40 (12): 2487–509.

Lees, L. (2003b) 'Visions of "urban renaissance": the Urban Task Force report and the Urban White Paper', in R. Imrie and M. Raco (eds), *Urban Renaissance? New Labour, Community and Urban Policy*. Bristol: Policy Press, pp. 61–82.

Lefebvre, H. (1991) *The Production of Space*. Oxford: Blackwell.

Lefebvre, H. (1996 [1968]) 'The Right to the City', in E. Kofman and E. Lebas (eds), *Writings on Cities*. London: Blackwell, pp. 147–59.

Lefebvre, H. (2003 [1970]) *The Urban Revolution*. Minneapolis, MN: University of Minnesota Press.

Lefebvre, H. (2006 [1961]) 'The Social Text', in S. Elden, E. Kofman and E. Lebas (eds), *Henri Lefebvre: Key Writings*. London: Bloomsbury, pp. 88–92.

Lefebvre, H. (2009) *State, Space, World*. Minneapolis, MN: University of Minnesota Press.

Lichtenstein, G. (1974) 'Stuntman, eluding guards, walks a tightrope between Trade Center Towers', *New York Times*, 8 August, p. 20.

Lipinski, J. (2012) 'Leaving his footprints on the city', *New York Times Online*, 23 March. http://www.nytimes.com/2012/03/25/nyregion/matt-greens-goal-is-to-walk-every-street-in-new-york-city.html (accessed 4 September 2014).

Lyndon, M., Bartman, D., Woudstra, R. and Khawazad, A. (2011) Tactical Urbanism Vol. 1: Short Term Action | | Long Term Change'. http://issuu.com/streetplanscollaborative/docs/tactical_urbanism_vol.1 (accessed 3 October 2013).

Lyndon, M., Bartman, D., Garcia, T., Preston, R. and Woudstra, R. (2012) 'Tactical Urbanism Vol. 2: Short Term Action | | Long Term Change'. http://issuu.com/streetplanscollaborative/docs/tactical_urbanism_vol_2_final (accessed 3 October 2013).

McAuliffe, C. (2012) 'Graffiti or street art? Negotiating the moral geographies of the creative city', *Journal of Urban Affairs*, 34 (2): 189–206.

McCalman, P. (2004) 'Foreign direct investment and intellectual property rights: evidence from Hollywood's global distribution of movies and videos', *Journal of International Economics*, 62 (1): 107–23.

McCann, E. and Ward, K. (2011) 'Introduction: urban assemblages: territories, relations, practices, and power', in E. McCann and K. Ward (eds), *Mobile Urbanism: Cities and Policymaking in the Global Age*. Minneapolis, MN: University of Minnesota Press, pp. xiii–xxxv.

McCarthy, J. (2005) 'Cultural quarters and regeneration: the case of Wolverhampton', *Planning, Practice and Research*, 20 (3): 297–311.

MacDonald, F. (2013) *The Popular History of Graffiti: From the Ancient World to the Present*. New York: Skyhorse.

McFarlane, C. (2011) 'The city as assemblage: dwelling and urban space', *Environment and Planning D*, 29 (4): 649–71.

McGuirk, J. (2014) *Radical Cities: Across Latin America in Search of a New Architecture*. London: Verso.

Machon, J. (2013) *Immersive Theatres: Intimacy and Immediacy in Contemporary Performance*. London: Palgrave Macmillan.

Mackay, R. (2011) '"Going backwards in time to talk about the present": man on wire and verticality after 9/11', *Comparative American Studies*, 9 (1): 3–20.

McQuire, S. (2008) *The Media City: Media, Architecture and Urban Space*. London: Sage.

Maffesoli, M. (1995) *The Time of the Tribes: The Decline of Individualism in Mass Society*. London: Sage.

Malanga, S. (2004) 'The curse of the creative class', *City Journal*, Winter. http://www.city-journal.org/html/14_1_the_curse.html (accessed 4 September 2014).

Marcuse, P. (2009) 'From critical urban theory to the right to the city', *City*, 12 (2–3): 195–7.

Markusen, A. (2006) 'Urban development and the politics of a creative class: evidence from a study of artists', *Environment and Planning A*, 38 (10): 1921–40.

Massey, D., Quintas, P. and Wield, D. (1992) *High-Tech Fantasies: Science Parks in Society, Science and Space*. London: Routledge.

Mayer, M. (2013) 'First world urban activism', *City: Analysis of Urban Trends, Culture, Theory, Policy, Action*, 17 (1): 5–19.

Metcalfe, J. (2014) 'Daredevil Russians climb the world's second-tallest skyscraper', *CityLab*, 13 February. http://www.citylab.com/design/2014/02/daredevil-russians-climb-worlds-second-tallest-skyscraper/8392/ (accessed 3 September 2014).

Minji, Y. (2013) 'Developer pulls down part of M50 graffiti wall', *ShanghaiDaily.com*, 13 December. http://www.shanghaidaily.com/metro/Developer-pulls-down-part-of-M50-graffiti-wall/shdaily.shtml (accessed 5 September 2014).

Minogue, E. (2014) 'This generation's "No Ball Games" sign ... The ever growing restriction and encroachment on space and place for you ...', 20 June, 8:25 p.m. (@eugeneminogue).

Mommaas, H. (2004) 'Cultural clusters and post-industrial city: towards the remapping of urban cultural policy', *Urban Studies*, 41 (3): 507–32.

Montgomery, J. (2003) 'Cultural Quarters as mechanisms for urban regeneration. Part 1: Conceptualising Cultural Quarters', *Planning Practice and Research*, 18 (4): 293–306.

Montgomery, J. (2008) *The New Wealth of Cities: City Dynamics and the Fifth Wave*. Aldershot: Ashgate.

Moore, M. and Prain, L. (2009) *Yarn Bombing: The Art of Crochet and Knit Graffiti*. Vancouver: Arsenal Pulp Press.

Moretti, E. (2012) *The New Geography of Jobs*. New York: Houghton Mifflin Harcourt.

Morley, P. (2013) 'Northern exposure: the BBC in Salford', *Financial Times Online*, 31 May. http://www.ft.com/cms/s/2/f5cc1646-c90b-11e2-9d2a-00144feab7de.html# axzz3CihcH8jP (accessed 8 September 2014).

Mott, C. and Roberts, S. M. (2014) 'Not everyone has (the) balls: urban exploration and the persistence of masculinist geography', *Antipode*, 46 (1): 229–45.

Mouffe, C. (2013) *Agonistics: Thinking the World Politically*. London: Verso.

Mould, O. (2009) 'Parkour, the city, the event', *Environment and Planning D: Society and Space*, 27: 738–50.

Mould, O. (2014a) 'Mediating Creative Cities: the role of planned media cities in the geographies of creative industry activity', in B. Derruder, S. Conventz, A. Thierstein and F. Witlox (eds), *Hub Cities in the Knowledge Economy: Seaports, Airports, Brainports*. Basingstoke: Ashgate, pp. 163–80.

Mould, O. (2014b) 'Tactical urbanism: the new vernacular of the Creative City', *Geography Compass*, 8 (8): 529–39.

Mould, O. (forthcoming) '"Jumped-up little ..." Parkour, activism and young people', in P. Kraftl and K. Nairn (eds), *Geographies of Children and Young People*. London: Springer.

Newman, O. (1996) *Creating Defensible Space*. New York: Diane Publishing.

Nieuwenhuys, C. (1974) 'New Babylon'. http://www.notbored.org/new-babylon.html (accessed 4 September 2014).

Ninjalicious (2005) *Access All Areas: A User's Guide to the Art of Urban Exploration*. Canada: Infilpress.

Norcliffe, G., Bassett, K. and Hoare, T. (1996) 'The emergence of postmodernism on the urban waterfront: geographical perspectives on changing relationships', *Journal of Transport Geography*, 4 (2): 123–34.

O'Brien, D. (2013) *Cultural Policy: Management, Value and Modernity in the Creative Industries*. London: Routledge.

O'Brien, M. (2007) *Beneath the Neon: Life and Death in the Tunnels of Las Vegas*. Las Vegas, NV: Huntington Press.

O'Connor, J. (2007) *The Cultural and Creative Industries: A Review of the Literature*. London: Arts Council England.

Oakley, K. (2004) 'Not so cool Britannia: the role of the creative industries in economic development', *International Journal of Cultural Studies*, 7 (1): 67–77.

Ong, A. (2006) *Neoliberalism as Exception: Mutations in Citizenship and Sovereignty*. London: Duke University Press.

Peck, J. (2001) 'Neoliberalizing states: thin policies/hard outcomes', *Progress in Human Geography*, 25 (3): 445–55.

Peck, J. (2005) 'Struggling with the creative class', *International Journal of Urban and Regional Research*, 29: 740–70.

Peck, J. and Tickell, A. (2002) 'Neoliberalizing space', *Antipode*, 34 (3): 380–404.

Peck, J. and Tickell, A. (2007) 'Conceptualizing neoliberalism, thinking Thatcherism, contesting neoliberalism', in H. Leitner, J. Peck and E. Sheppard (eds), *Contesting Neoliberalism: Urban Frontiers*. New York: Guilford Press, pp. 26–50.

Peck, J., Theodore, N. and Brenner, N. (2013) 'Neoliberal urbanism redux', *International Journal of Urban and Regional Research*, 37 (3): 1091–9.

Peel (n.d.) 'Liverpool Waters'. http://peelez.co.uk/content/liverpoolwaters.php (accessed 4 September 2014).

People's Republic of Stokes Croft (2011) 'Mission Statement'. http://www.prsc.org.uk/mission.htm (accessed 20 August 2013).

Petit, P. (2008) *Man on Wire*. London: Skyhorse.

Pinder, D. (2005) *Visions of the City: Utopianism, Power and Politics in Twentieth-Century Urbanism*. Edinburgh: Edinburgh University Press.

Plant, S. (1992) *The Most Radical Gesture. The Situationist International in a Postmodern Age*. London: Routledge.

Plaza, B., Tironi, M. and Haarich, S. (2009) 'Bilbao's art scene and the "Guggenheim effect" revisited', *European Planning Studies*, 17 (11): 1711–29.

Poe, E. A. (1840) 'The man of the crowd', *Burton's Gentlemen's Magazine*, December.

Porter, T. (1995) *Trust in Numbers: The Pursuit of Objectivity in Science and Public Life*. Princeton, NJ: Princeton University Press.

Pratt, A. (2004) 'Creative clusters: towards the governance of the creative industries production system?', *Media International Australia*, 112: 50–66.

Pratt, A. (2008) 'Creative cities: the cultural industries and the creative class', *Geografiska Annaler: Series B, Human Geography*, 90 (2): 107–17.

ProjektsMCR (2013) 'Opening Times and Prices'. http://www.projektsmcr.com/opening-times-and-prices (accessed 5 September 2014).

Proud, A. (2014) 'Why this "Shoreditchification" of London must stop', *Telegraph Online*, 13 January. http://www.telegraph.co.uk/men/thinking-man/10561607/Why-this-Shoreditchification-of-London-must-stop.html (accessed 5 September 2014).

Purcell, M. (2002) 'Excavating Lefebvre: the right to the city and its urban politics of the inhabitant', *GeoJournal*, 58 (2–3): 99–108.

Purcell, M. (2013) *The Down-Deep Delight of Democracy*. London: Wiley-Blackwell.

Rantisi, N. M. and Leslie, D. (2006) 'Branding the design metropole: the case of Montréal, Canada', *Area*, 38 (4): 364–76.

Robinson, J. (2002) 'Global and world cities: a view from off the map', *International Journal of Urban and Regional Research*, 26 (3): 531–54.

Roodhouse, S. (2006) *Cultural Quarters: Principles and Practice*. Bristol: Intellect.

Rosen, S. (1979) 'Wage-based indexes of urban quality of life', in P. Mieszkowski and M. Straszheim (eds), *Current Issues in Urban Economics*. Baltimore, MD: Johns Hopkins University Press, pp. 74–104.

Rosenberg, L. (2011) 'Art in Vacant Storefronts: A New Arena for Creative Research and Development'. PhD thesis, School of the Art Institute of Chicago.

Ross, A. (2008) 'The new geography of work: power to the precarious?', *Theory, Culture and Society*, 25 (7–8): 31–49.

Routledge, P. (2010) 'Sensuous solidarities: emotion, politics and performance in the Clandestine Insurgent Rebel Clown Army', *Antipode*, 44 (2): 428–52.

Santagata, W. (2002) 'Cultural districts, property rights and sustainable economic growth', *International Journal of Urban and Regional Research*, 26 (1): 9–23.

Sassen, S. (2001) *The Global City: New York, London, Tokyo*, revised edition. Princeton, NJ: University Press.

Scott, A. (2014) 'Beyond the creative city: cognitive–cultural capitalism and the new urbanism', *Regional Studies*, 48 (4): 565–578.

SE11 Action Team (2013) 'Local investment projects from the Beaufoy sale announced!' http://se11actionteam.blogspot.co.uk/2013/11/local-investment-projects-from-beaufoy.html (accessed 5 September 2014).

Shaw, K. (2013) 'Independent creative subcultures and why they matter', *International Journal of Cultural Policy*, 19 (3): 333–52.

Sklair, L. (2001) *Transnational Capitalist Class*. London: Blackwell.

Slater, T. (2011) 'Gentrification of the city', in G. Bridge and S. Watson (eds), *The New Blackwell Companion to the City*. London: Wiley, pp. 571–85.

Slater, T. (2013) 'Your life chances affect where you live: a critique of the "cottage industry" of neighbourhood effects research', *International Journal of Urban and Regional Research*, 37 (2): 367–87.

Smith, A. and von Krogh Strand, I. (2011) 'Oslo's new Opera House: cultural flagship, regeneration tool or destination icon?', *European Urban and Regional Studies*, 18 (1): 93–110.

Smith, M. (2013) 'Liverpool Waters redevelopment gets government green light', *Guardian Online*. http://www.theguardian.com/uk/2013/mar/05/liverpool-waters-redevelopment-green-light (accessed 4 September 2014).

Smith, N. (1987) 'Gentrification and the rent gap', *Annals of the Association of American Geographers*, 77 (3): 462–5.

Smith, N. (1996) *The New Urban Frontier: Gentrification and the Revanchist City*. New York: Psychology Press.

Smith, R. (2013) 'The ordinary city trap', *Environment and Planning A*, 45: 2290–304.

Smith, R. G. (2014) 'Beyond the global city concept and the myth of "command and control"', *International Journal of Urban and Regional Research*, 38 (1): 98–115.

Smyth, H. (2005) *Marketing the City: The Role of Flagship Developments in Urban Regeneration*. London: Taylor & Francis.

Snyder, G. J. (2009) *Graffiti Lives: Beyond the Tag in New York's Urban Underground*. New York: NYU Press.

Southbank Centre (2013a) Press release, 6 March: 'Southbank Centre unveils plans to transform Festival Wing and create world-class cultural centre in London'. http://www.southbankcentre.co.uk/sites/default/files/press_releases/southbank_centre_unveils_plans_for_festival_wing_transformation_press_release.pdf (accessed 19 November 2014).

Southbank Centre (2013b) 'The Festival Wing: Explore our Plans'. http://www.thefestivalwing.co.uk/explore-our-plans/ (accessed 5 September 2014).

Spinney, J. (2010) 'Performing resistance? Re-reading practices of urban cycling on London's South Bank', *Environment and Planning A*, 42 (12): 2914–937.

Springer, S. (2010) 'Neoliberalism and geography: expansions, variegations, formations', *Geography Compass*, 4 (8): 1025–38.

Storper, M. (2010) 'Why does a city grow? Specialisation, human capital or institutions?', *Urban Studies*, 47 (10): 2027–50.

Storper, M. and Scott, A. J. (2009) 'Rethinking human capital, creativity and urban growth', *Journal of Economic Geography*, 9: 147–67.

Streets Plan Collaborative (n.d.) http://www.streetplans.org (accessed 19 November 2014).

Swyngedouw, E. (2002) 'The strange respectability of the situationist city in the society of the spectacle', *International Journal of Urban and Regional Research*, 26 (1): 153–65.

Taylor, P. (2004) *World City Network*. London: Routledge.

Taylor, P. J. (2013) *Extraordinary Cities: Millennia of Moral Syndromes, World-Systems and City/State Relations*. London: Edward Elgar.

Taylor, P., Ni, P., Derudder, B., Hoyler, M., Witlox, F. and Huang, J. (2010) 'Command and control centres in the global economy', in P. Taylor, P. Ni, B. Derudder, M. Hoyler, F. Witlox and J. Huang (eds), *Global Urban Analysis: A Survey of Cities in Globalisation*. London: Earthscan, pp. 17–21.

Tel Aviv Mayor's Office (2010) 'Tel Aviv Global and Tourism'. http://www.tel-aviv.gov. il/eng/GlobalCity/Pages/GlobalCityLoby.aspx?tm=24 (accessed 3 September 2014).

Thomas, A. (2010) *Prague Palimpsest: Writing, Memory, and the City.* Chicago: University of Chicago Press.

Thorpe, H. and Ahmad, N. (2013) 'Youth, action sports and political agency in the Middle East: lessons from a grassroots parkour group in Gaza', *International Review for the Sociology of Sport.* DOI: 1012690213490521 (early access online 18 June 2013).

Time Out Shanghai (2011) 'The end of the M50 graffiti wall', *Time Out Shanghai Online,* 26 August. http://www.timeoutshanghai.com/features/Around_Town-Around_ Town/3951/The-end-of-the-M50-graffiti-wall.html (accessed 5 September 2014).

Turok, I. (2003) 'Cities, clusters and creative industries: the case of film and television in Scotland', *European Planning Studies*, 11 (5): 549–65.

Vanolo, A. (2008) 'The image of the creative city: some reflections on urban branding in Turin', *Cities*, 25 (6): 370–82.

Vanolo, A. (2013) 'Alternative capitalism and creative economy: the case of Christiania', *International Journal of Urban and Regional Research*, 37 (5): 1785–98.

Vickery, J. (2015) *Creative Cities and Public Cultures: Art, Democracy and Urban Lives.* London: Routledge.

Virilio, P. (1994) *The Vision Machine*, trans. J. Rose. London: British Film Institute.

Visit Leicester (2013) 'Cultural Quarter Leicester'. http://www.visitleicester.info/things-to-see-and-do/cultural-quarter/ (accessed 4 September 2014).

Waclawek, A. (2011) *Graffiti and Street Art.* London: Thames & Hudson.

Wallace, J. (2013) 'Yarn bombing, knit graffiti and underground brigades: a study of craftivism and mobility', *Journal of Mobile Media: Sound Moves*, 7 (1). http://wi. mobilities.ca/yarn-bombing-knit-graffiti-and-underground-brigades-a-study-of-craftivism-and-mobility/ (accessed 5 September 2014).

Wallerstein, I. M. (2004) *World-Systems Analysis: An Introduction.* Durham, NC: Duke University Press.

Watson, V. (2009) '"The planned city sweeps the poor away…": urban planning and 21st century urbanisation', *Progress in Planning*, 72: 151–93.

WebUrbanist (2014) 'Bricksy: 20 classic Banksy street artworks recreated in LEGO', *WebUrbanist.* http://weburbanist.com/2014/06/14/bricksy-20-classic-banksy-street-artworks-recreated-in-lego/ (accessed 5 September 2014).

Wenger, E. (1998) *Communities of Practice: Learning, Meaning, and Identity.* Cambridge: Cambridge University Press.

Whitford, M. (1992) *Getting Rid of Graffiti: A Practical Guide to Graffiti Removal and Anti-graffiti Protection.* London: Taylor & Francis.

Wigley, M. (1998) *Constant's New Babylon: The Hyper-architecture of Desire.* London: 010 Publishers.

Wilson, D. (2004) 'Toward a contingent urban neoliberalism', *Urban Geography*, 25: 771–83.

Wilson, D. and Keil, R. (2008) 'The real creative class', *Social and Cultural Geography*, 9 (8): 841–7.

Wirth, L. (1938) 'Urbanism as a way of life', *American Journal of Sociology*, 44: 1–24.

Wood, D. (2002) 'Foucault and panopticism revisited', *Surveillance and Society*, 1 (3): 234–9.

Wood, Z. (2014) 'Britain has world's most billionaires per capita', *Guardian Online,* 11 May. http://www.theguardian.com/business/2014/may/11/britain-worlds-most-billionaires-per-capita (accessed 5 September 2014).

Zheng, J. (2011) '"Creative industry clusters" and the "entrepreneurial city" of Shanghai', *Urban Studies*, 48 (16): 3561–82.

Zimmerman, J. (2008) 'From brew town to cool town: neoliberalism and the creative city development strategy in Milwaukee', *Cities*, 25 (4): 230–42.

Zukin, S. (2010) *Naked City: The Death and Life of Authentic Urban Places*. Oxford: Oxford University Press.

Films

Exit Through the Gift Shop (2010) [Film] dir. Banksy. UK: Paranoid Pictures.

Man on Wire (2008) [Film] dir. James Marsh. UK: Discovery Films.

Index